Springer Advanced Texts in Chemistry

Springer Advanced Texts in Chemistry

Series Editor: Charles R. Cantor

Jan Drenth

Principles of Protein
X-ray Crystallography

With 141 Illustrations

Springer-Verlag
New York Berlin Heidelberg London Paris
Tokyo Hong Kong Barcelona Budapest

Jan Drenth
Laboratory of Biophysical Chemistry
Nijenborgh 4
9747 AG Groningen
The Netherlands

Series Editor:
Charles R. Cantor
Boston University
Center for Advanced Research in Biotechnology
Boston, MA 02215, USA

Cover illustrations courtesy of Adrian R. Ferré-D'Amaré and Stephen K. Burley, The Rockefeller University.
Foreground, front cover: The four-α-helix bundle moiety of transcription factor Max. Reproduced with permission from *Nature* 363:38–45 (1993). **Background**: Electron density map of the DNA-binding domain of the transcription factor Max.

Library of Congress Cataloging-in-Publication Data
Drenth, Jan.
    Principles of protein X-ray crystallography / Jan Drenth.
        p.   cm. — (Springer advanced texts in chemistry)
    Includes bibliographical references and index.
    ISBN 0-387-94091-X
        1. Proteins—Analysis.   2. X-ray crystallography.   I. Title.
    II. Series.
    QP551.D74   1994
    574.19′245—dc20   93-6235

Printed on acid-free paper.

Acquiring Editor: Robert Garber.
Production coordinated by Chernow Editorial Services, Inc., and managed by Ellen Seham; manufacturing supervised by Vincent Scelta.
Typeset by Best-set Typesetter Ltd., Hong Kong.
Printed and bound by R.R. Donnelley and Sons, Harrisonburg, VA.
Printed in the United States of America.

9 8 7 6 5 4 3 2 1

ISBN 0-387-94091-X Springer-Verlag New York Berlin Heidelberg
ISBN 3-540-94091-X Springer-Verlag Berlin Heidelberg New York

Series Preface

New textbooks at all levels of chemistry appear with great regularity. Some fields like basic biochemistry, organic reaction mechanisms, and chemical thermodynamics are well represented by many excellent texts, and new or revised editions are published sufficiently often to keep up with progress in research. However, some areas of chemistry, especially many of those taught at the graduate level, suffer from a real lack of up-to-date textbooks. The most serious needs occur in fields that are rapidly changing. Textbooks in these subjects usually have to be written by scientists actually involved in the research which is advancing the field. It is not often easy to persuade such individuals so set time aside to help spread the knowledge they have accumulated. Our goal, in this series, is to pinpoint areas of chemistry where recent progress has outpaced what is covered in any available textbooks, and then seek out and persuade experts in these fields to produce relatively concise but instructive introductions to their fields. These should serve the needs of one semester or one quarter graduate courses in chemistry and biochemistry. In some cases, the availability of texts in active research areas should help stimulate the creation of new courses.

Charles R. Cantor

Preface

Macromolecules are the principal nonaqueous components of living cells. Among the macromolecules (proteins, nucleic acids, and carbohydrates), proteins are the largest group. Enzymes are the most diverse class of proteins because nearly every chemical reaction in a cell requires a specific enzyme. To understand cellular processes, knowledge of the three-dimensional structure of enzymes and other macromolecules is vital. Two techniques are widely used for the structural determination of macromolecules at atomic resolution: X-ray diffraction of crystals and nuclear magnetic resonance (NMR). While NMR does not require crystals and provides more detailed information on the dynamics of the molecule in question, it can be used only for biopolymers with a molecular weight of less than 20,000. X-ray crystallography can be applied to compounds with molecular weight up to at least 10^6. For many proteins, the difference is decisive in favor of X-ray diffraction.

The pioneering work by Perutz and Kendrew on the structure of hemoglobin and myoglobin in the 1950s led to a slow but stready increase in the number of proteins whose structure was determined using X-ray diffraction. The introduction of sophisticated computer hardware and software dramatically reduced the time required to determine a structure while increasing the accuracy of the results. In recent years, recombinant DNA technology has further stimulated interest in protein structure determination. A protein that was difficult to isolate in sufficient quantities from its natural source can often be produced in arbitrarily large amounts using expression of its cloned gene in a microorganism. Also, a protein modified by site-directed mutagenesis of its gene can be created for scientific investigation and industrial application. Here, X-ray diffraction plays a crucial role in guiding the molecular biologist to the best amino acid positions for modification. Moreover, it is often important to learn what effect a change in a protein's sequence will have on its three-dimensional structure. Chemical and pharmaceutical companies have become very active in the field of protein structure determination because of their interest in protein and drug design.

This book presents the principles of the X-ray diffraction method. Although I will discuss *protein* X-ray crystallography exclusively, the same techniques can in principle be applied to other types of macro-molecules and macromolecular complexes. The book is intended to serve both as a textbook for the student learning crystallography, and as a reference for the practicing scientist. It presupposes a familiarity with mathematics at the level of upper level undergraduates in chemistry and biology, and is designed for the researcher in cell and molecular biology, biochemistry, or biophysics who has a need to understand the basis for crystallographic determination of a protein structure.

I would like to thank the many colleagues who have read the manuscript and have given valuable comments, especially Aafje Vos, Shekhar and Sharmila Mande, Boris Strokogrytov, and Risto Lagratto.

Contents

Chapter 1
Crystallizing a Protein

1.1. Introduction

Students new to the protein X-ray crystallography laboratory may understandably be confused when colleagues discuss Fouriers and Pattersons or molecular replacement and molecular dynamics refinement. However, they understand immediately that the first requirement for protein structure determination is to grow suitable crystals. Without crystals there can be no X-ray structure determination of a protein! In this chapter we discuss the principles of protein crystal growth and as an exercise give the recipe for crystallizing the enzyme lysozyme. We shall also generate an X-ray diffraction picture of a lysozyme crystal. This will provide an introduction to X-ray diffraction. The chapter concludes with a discussion of the problems encountered.

1.2. Principles of Protein Crystallization

Obtaining suitable single crystals is the least understood step in the X-ray structural analysis of a protein. The science of protein crystallization is an underdeveloped area, although interest is growing, spurred especially by microgravity experiments in space flights. Protein crystallization is mainly a trial-and-error procedure in which the protein is slowly precipitated from its solution. The presence of impurities, crystallization nuclei, and other unknown factors plays a role in this process. As a general rule, however, the purer the protein, the better the chances to grow crystals. The purity requirements of the protein crystallographer are different and more stringent than the requirements of the biochemist, who would be satisfied if, for example, the catalytic activity of an enzyme is sufficiently high. On the other hand, to achieve protein crystallization not only should other compounds be absent, but all molecules of the protein should have the same surface properties, especially the same charge distribution on their surface, since this influences the packing of the molecules in the crystal.

The crystallization of proteins involves four important steps:

1. The purity of the protein is determined. If it is not extremely pure, further purification will generally be necessary to achieve crystallization.

2. The protein is dissolved in a suitable solvent from which it must be precipitated in crystalline form. The solvent is usually a water–buffer solution, sometimes with an organic solvent such as 2-methyl-2,4-pentanediol (MPD) added. Normally, the precipitant solution is also added, but only to such a concentration that a precipitate does not develop. Membrane proteins, which are insoluble in a water–buffer or a water–organic solvent, require in addition a detergent.

3. The solution is brought to supersaturation. In this step small aggregates are formed, which are the nuclei for crystal growth. For the crystallization of small molecules, which is much better understood than the crystallization of proteins, the spontaneous formation of nuclei requires a supply of surface tension energy. Once the energy barrier has been passed, crystal growth begins. The energy barrier is easier to overcome at a higher level of supersaturation. Therefore, spontaneous formation of nuclei is best achieved at a high supersaturation. We assume that this is also true for the crystallization of proteins. Formation of nuclei can be studied as a function of supersaturation and other parameters by a number of techniques, including light scattering, fluorescence depolarization, and electron microscopy.

4. Once nuclei have formed, actual crystal growth can begin. As for low-molecular-weight compounds, the attachment of new molecules to the surface of a growing crystal occurs at steps on the surface. This is because the binding energy is larger at such positions than if the molecule attaches to a flat surface. These steps are either created by defects in the crystalline order or occur at nuclei formed randomly on the surface.

To achieve crystal growth, supersaturation must be reduced to a lower level; maintaining a high supersaturation would result in the formation of too many nuclei and therefore too many small crystals. Also, crystals should grow slowly to reach a maximum degree of order in their structure. In practice, however, this fundamental rule is not always obeyed. The easiest way to change the degree of supersaturation is by changing the temperature (Figure 1.1).

Precipitation of the protein can be achieved in more than one way. A common precipitation method involves increasing the effective concentration of the protein, usually by adding a salt (salting-out) or

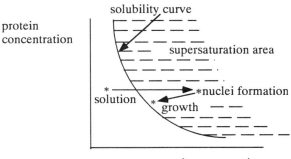

Figure 1.1. A typical solubility curve for a protein, as a function of the salt concentration or another parameter. For best results the crystals should be grown at a lower level of supersaturation than is required for formation of nuclei.

polyethyleneglycol (PEG). Either serves to immobilize water, thereby increasing the effective protein concentration. The most popular salt is ammonium sulfate because of its high solubility.

Some proteins are poorly soluble in pure water but do dissolve if a small amount of salt is added. By removing the salt, the protein precipitates. This "salting-in" effect is explained by regarding the protein as an ionic compound. According to the Debye–Hückel theory for ionic solutions, an increase in the ionic strength lowers the activity of the ions in the solution and increases the solubility of ionic compounds. Alternatively, one can regard salting-in as the result of a competition between charged groups on the surface of the protein molecule and the ions in the solution. In the absence of solvent ions the protein precipitates by Coulomb attraction between opposite charges on different protein molecules. If ions are added they screen the charged groups on the protein and increase the solubility. A second method of protein precipitation is to diminish the repulsive forces between the protein molecules or to increase the attractive forces. These forces are of different types: electrostatic, hydrophobic, and hydrogen bonding. Electrostatic forces are influenced by an organic solvent such as alcohol, or by a change in pH. The strength of hydrophobic interactions increases with temperature.

To summarize, the usual procedure for crystallizing a protein is to:

1. check the purity carefully;
2a. slowly increase the concentration of the precipitant such as PEG, salt, or an organic solvent; or
2b. change the pH or the temperature.

In practice the amount of protein available for crystallization experiments is often very small. To determine the best crystallization conditions it is usually necessary to carry out a great number of experiments; hence a minimum amount of protein should be used per experiment. A single protein crystal of reasonable size ($0.3 \times 0.3 \times 0.3\,mm = 0.027\,mm^3$) weighs approximately $15\,\mu g$. Therefore, $1\,mg$ of purified protein is sufficient to perform about 65 crystallization experiments.

1.3. Crystallization Techniques

1.3.1. Batch Crystallization

This is the oldest and simplest method for protein crystallization. The principle is that the precipitating reagent is instantaneously added to a protein solution, suddenly bringing the solution to a state of high supersaturation. With luck crystals grow gradually from the supersaturated solution without further processing. An automated system for microbatch crystallization has been designed by Chayen et al. (1990, 1992).

1.3.2. Liquid–Liquid Diffusion

In this method the protein solution and the solution containing the precipitant are layered on top of each other in a small-bore capillary; a melting point capillary can conveniently be used (Figure 1.2). The lower layer is the solution with higher density (for example, a concentrated ammonium sulfate or PEG solution). If an organic solvent such as MPD is used as precipitant, it forms the upper layer. For a 1:1 mixture the concentration of the precipitant should be two times its desired final concentration. The two solutions (approximately $5\,\mu l$ of each) are introduced into the capillary with a syringe needle, beginning with the

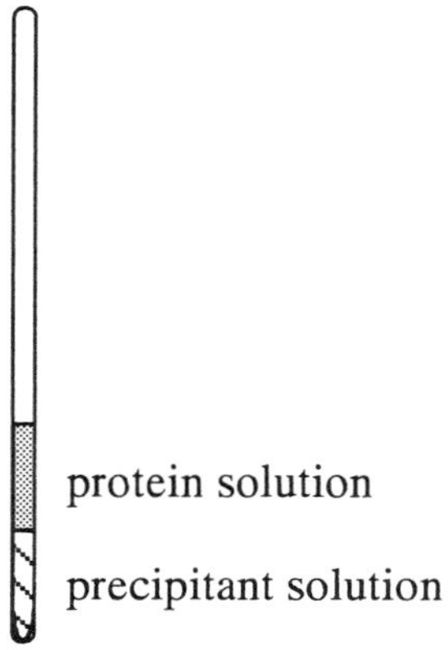

protein solution

precipitant solution

Figure 1.2. Liquid–liquid diffusion in a melting point capillary. If the precipitant solution is the denser one, it forms the lower layer.

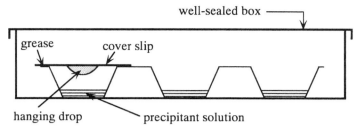

Figure 1.3. The hanging drop method of protein crystallization. Using a tray with depressions, the protein solution is suspended as a drop from a glass cover slip above the precipitant solution in a sealed depression. The glass slip is siliconized to prevent spreading of the drop. Equilibrium is reached by diffusion of vapor from the drop to the precipitating solution or vice versa. All of the depressions in the tray can of course be used.

lower one. Spinning in a simple swing-out centrifuge removes air bubbles. The upper layer is added and a sharp boundary is formed between the two layers. They gradually diffuse into each other.

1.3.3. Vapor Diffusion

1.3.3.1. By the Hanging Drop Method

In this method drops are prepared on a siliconized microscope glass cover slip by mixing 3–10 µl of protein solution with the same volume of precipitant solution. The slip is placed upside down over a depression in a tray; the depression is partly filled with the required precipitant solution (approximately 1 ml). The chamber is sealed by applying oil or grease to the circumference of the depression before the cover slip is put into place (Figure 1.3).

1.3.3.2. Sitting Drop

If the protein solution has a low surface tension, it tends to spread out over the cover slip in the hanging drop method. In such cases the sitting drop method is preferable. A schematic diagram of a sitting drop vessel is shown in Figure 1.4.

1.3.4. Dialysis

As with the other methods for achieving protein crystallization, many variations of dialysis techniques exist. The advantage of dialysis is that the precipitating solution can be easily changed. For moderate amounts of protein solution (more than 0.1 ml), dialysis tubes can be used, as shown

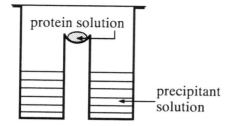

Figure 1.4. The sitting drop method for performing protein crystallization.

in Figure 1.5a. The dialysis membrane is attached to a tube by means of a rubber ring. The membrane should be rinsed extensively with water before use or, preferably, boiled in water for about 10 mins. For microliter amounts of protein solution, one may use either a thick-walled micro-capillary (Zeppezauer method) or a plexiglass "button" covered with a dialysis membrane (Figure 1.5b). The disadvantage of the button is that any protein crystals in the button cannot be observed with a polarizing microscope.

Another microdialysis procedure is illustrated in Figure 1.6. Five microliters of protein solution is injected into a capillary, which is covered by a dialysis membrane. The membrane may be fastened with a piece of tubing. The protein solution is spun down in a simple centrifuge and the capillary closed with modeling clay. The capillary is then placed in an Eppendorf tube containing the dialysis solution.

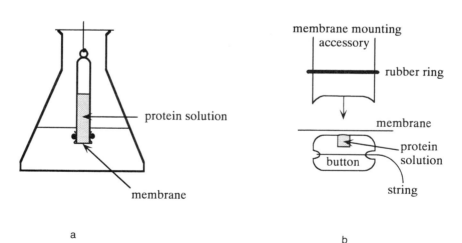

Figure 1.5. Protein crystallization by dialysis. If a relatively large amount of protein is available, dialysis can be performed as in (a). Smaller amounts can be crystallized in a button (b).

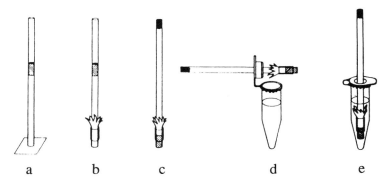

Figure 1.6. Protein crystallization by microdialysis procedure. In (a) and (b) the protein solution is still somewhere in the capillary. In (c) it has been spun to the bottom of the tube and is in contact with the membrane. In (d) and (e) the capillary is mounted in an Eppendorf tube containing dialysis solution.

Crystallization of a protein is a multiparameter problem in which the parameters are varied in the search for optimal crystallization conditions. The most common parameters that are changed include protein concentration, the nature and concentration of the precipitant, pH, and temperature. Specific additives that affect the crystallization can also be added in low concentration. An appropriate statistical design to use for this problem is the factorial or incomplete factorial method, in which a table is constructed with a number of values entered for each parameter (e.g., pH set at 5, 6, and 7). The table is constructed to take into account reasonable boundary conditions. After approximate crystallization conditions have been found, the conditions can be optimized (Carter, 1990).

McPherson (1992) described a method that is in between pure trial and error and the factorial design approach for identifying initial crystallization conditions. However, in his method a large number of conditions must still be tried (Table 1.1).

Robotic workstations can be highly effective in conducting such multi-factorial experiments (an example is found in Chayen et al., 1990).[1]

1.4. Crystallization of Lysozyme

After learning the principles of protein crystallization, it is time to do a crystallization experiment. The most convenient protein to start with is hen egg white lysozyme. It can be obtained commercially in pure form, is

---

[1] Protein crystallization is extensively discussed in: Ducruix, A. and Giegé, R. (1992). For the crystallization of membrane proteins, see Michel, H. (1990).

Table 1.1. Screening Reagents[a]

Box 1 and 2 buffers	Box 4 salt solutions
1. pH 4.0 0.1 M citrate or acetate	1. 50% sat. sodium phosphate, pH 7.0
2. pH 5.0 0.1 M acetate	2. 50% sat. magnesium sulfate
3. pH 5.5 0.1 M MES or cacodylate	3. 50% sat. sodium citrate pH, 7.0
4. pH 6.0 0.1 M MES or cacodylate or phosphate	4. 50% sat. lithium sulfate
5. pH 6.5 0.1 M cacodylate or phosphate	5. 50% sat. ammonium phosphate, pH 7.0
6. pH 7.0 0.1 M phosphate	6. 50% sat. sodium acetate
7. pH 7.5 0.1 M Tris-HCl	7. 50% sat. sodium chloride
8. pH 8.0 0.1 M HEPES or Tris	8. 50% sat. calcium sulfate
9. pH 8.5 0.1 M Tris or glycine	9. 50% sat. ammonium formate
Droplet = 5 µl protein + 5 µl buffer + 5 µl reservoir	Droplet = 5 µl protein + 5 µl salt + 5 µl water
Reservoir Box 1 = 14% PEG 3350	Reservoir = 45% sat. ammonium sulfate
Reservoir Box 2 = 45% sat. ammonium sulfate	

Box 3 nonvolatile organics	Cryschem or Linbro tray
1. 16.5% polyeneamine	1. *n*-Propanol
2. 50% v/v PEG 400	2. Isopropanol
3. 15% w/v PEG 20,000	3. Dioxane
4. 30% v/v Jeffamine ED 4000	4. Ethanol
5. 50% sat. PEG 1000 monostearate	5. *tert*-Butanol
6. 30% PEG 1000	6. DMSO
7. 30% Jeffamine ED 2001	
8. 50% propylene glycol	
9. 60% hexanediol	
Droplet = 5 µl protein + 5 µl organic + 5 µl water	Droplet = 5 µl protein + 5 µl water
Reservoir = 45% sat. ammonium sulfate	Reservoirs 0.75 ml of 35% organic solvent
Alternates: Methyl pentanediol 30% Jeffamine ED-900 30% Jeffamine ED-600	Alternates: Methanol Acetone *n*-Butanol

[a] From McPherson (1992), with permission.

relatively cheap, and can be used immediately for a crystallization experiment. In addition to native lysozyme, a heavy atom derivative of the protein will be crystallized, because the determination of a protein structure sometimes requires X-ray diffraction patterns of crystals of the native protein as well as of one or more heavy-atom-containing derivatives. In this experiment the mercury-containing reagent *p*-chloromercuriphenyl sulfonate is used.

1. Prepare a sodium acetate buffer solution of pH 4.7 by dissolving
   a. 1.361 g of sodium acetate in 50 ml water (purified by reverse osmosis or double distillation);
   b. 0.572 ml glacial acetic acid in 50 ml water.
   The acetic acid solution is added to the salt solution until the pH reaches 4.7.
2. Prepare a precipitant solution of sodium chloride
   a. for the native crystals by preparing 30 ml of 10% (w/v) NaCl in the sodium acetate buffer;
   b. for the derivative crystals by dissolving 0.041 g of the sodium salt of *p*-chloromercuriphenyl sulfonate (PCMS) in 5 ml of the precipitant solution prepared in 2a.

Warning! Heavy atom reagents can be very toxic or radioactive. Handle them with care and wear gloves. Discard excess reagent and solution in a special container.

---

3. Dissolve the native lysozyme in the acetate buffer to a concentration of 50 mg/ml (e.g., 10 mg in 200 µl). The heavy-atom derivative is less soluble and its concentration should by only 15 mg/ml in the same buffer. To remove insoluble particles the solutions are centrifuged at 10,000 rpm and 4°C for 10 min prior to setting up the crystallization experiments.
4. Take two trays, one for the native and one for the derivative protein, with at least 5 × 3 depressions in each (Figure 1.7). The crystallization

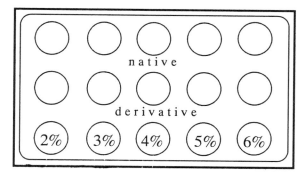

Figure 1.7. A tray for the crystallization of lysozyme. Five experiments each for the native protein and for the heavy atom derivative can be performed with increasing concentrations of the precipitant solution: 2–6% NaCl.

experiments are performed with five concentrations of the precipitant solution: 2, 3, 4, 5, and 6% NaCl and two replicates per condition. Use the extra row of depressions to label the experimental conditions. Fill each depression with 1 ml of the precipitant solution at the required concentration. Prepare the hanging drops on the siliconized side of microscope cover slips: pipette 10 µl of the enzyme solution onto a cover slip and mix with 10 µl of the precipitant solution. Brush a film of immersion oil around each depression on the surface of the tray to seal (Figure 1.3).

The lysozyme crystals will grow in a day or so and you will probably be anxious to see how your crystals will diffract X-rays. However, we will first consider some general features about crystals.

1.5. A Preliminary Note on Crystals

Crystals occur in a great variety of shapes and colors and naturally grown crystals have been used as gems since prehistoric times. Their flat faces reflect the regular packing of the molecules, atoms, or ions in the crystal. Crystals are distinguished from amorphous substances by their flat faces and by their anisotropy: some of their physical properties are dependent on the direction of measurement in the crystal. For some properties crystals may be isotropic, but usually not for all of them. You can easily observe the anisotropy of the lysozyme crystals you have just grown by examining them between crossed polarizers in a polarizing microscope. They display beautiful colors because the crystals are birefringent: they have two different refraction indices. If you rotate the microscope platform holding the crystals, they will become completely dark at a certain position and every 90° away from it; extinction thus occurs four times in a complete revolution.

The flat faces and anisotropy of crystals reflect their regular packing of molecules, atoms, or ions. This regular arrangement cannot be observed with the naked eye or with a light microscope because the particles are too small, but can be visualized with an electron microscope (Figure 1.8). If the resolution of the microscope would be high enough, the atomic structure inside the large protein molecules could be observed. Unfortunately this is not possible because biological substances can be observed in an electron microscope with a resolution of only 10–20 Å, due to limitations in specimen preparation. Although atomic resolution cannot be reached in this way, electron micrographs of crystals show convincingly the regular packing of molecules in the crystals. It is because of this regular arrangement that the crystals diffract X-rays.

Before starting the X-ray diffraction experiment it can be illuminating to demonstrate diffraction with visible light and a grating. The principle is

Figure 1.8. An electron micrograph of a crystal face of the oxygen transporting protein hemocyanin from *Panulirus interruptus*. The molecular weight of this protein is 450,000; magnification: 250,000×. Courtesy of E.F.J. van Bruggen.

the same, the only difference being that the grating is two-dimensional and the crystal is three-dimensional. Take a laser pointer as light source and ask an electron microscopist for one of the copper grids on which specimens are mounted. In a dark room you will see a beautiful pattern of diffraction spots. Rotation of the grid causes the pattern to rotate with it. If you can borrow grids with different spacings, the pattern from the grid with larger spacing will have the diffraction spots closer together: here you observe "reciprocity" between the diffraction pattern and the grid. Alternatively, you can perform the experiment with transparent woven fabric. Note that stretching the fabric horizontally causes the diffraction pattern to shrink horizontally.

1.6. Preparation for an X-ray Diffraction Experiment

You should now perform the real X-ray diffraction experiment using one of the crystals you have just grown. However, you should be aware of an important difference between crystals of small molecules and crystals of a protein. In protein crystals, the spherical or egg-shaped molecules are loosely packed with large solvent-filled holes and channels, which normally occupy 40–60% of the crystal volume. This is an advantage for the reaction of the protein with small reagent molecules; they can diffuse

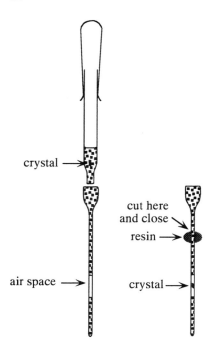

Figure 1.9. Mounting a protein crystal in a glass capillary between layers of mother liquor.

crystal →

cut here and close ↘

resin →

air space →

crystal →

through these channels and reach reactive sites on all protein molecules in the crystal. However, the high solvent content causes problems in handling the crystals because removal of the solvent destabilizes the crystal. Therefore, protein crystals should always be kept in their mother liquor or in the saturated vapor of their mother liquor, even when exposed to X-rays. For that purpose they are mounted in thin-walled capillaries of borosilicate glass or quartz. It requires a little practice to be able to work with protein crystals and to mount them as shown in Figure 1.9.

The procedure is as follows. Fill the X-ray capillary with the mother liquor in which the crystals were grown or with the precipitation solution in the bottom of the tray, but leave an air space where the crystal should be. Suck up one of the crystals together with mother liquor, let it descend to the tip of the pipet, and then touch the pipet tip to the liquid in the capillary. The crystal continues its way down until it reaches the air space. Push it carefully into the open space with a thin glass fiber. Excess liquid must be removed and the capillary closed. This can be done with a resin that can be melted using a soldering iron. Finally, for handling and for mounting on the X-ray camera, a piece of modeling clay is wrapped around the resin. Lysozyme crystals have a 4-fold symmetry axis and for the diffraction experiment they should be mounted with this axis perpendicular to the axis of the capillary, as seen in Figure 1.10. This has

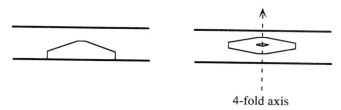

4-fold axis

Figure 1.10. Two perpendicular views of a lysozyme crystal mounted in a capillary. The size of the crystal is 0.2–0.5 mm.

the advantage that in the diffraction experiment the symmetry axis can be aligned parallel to the X-ray beam.

Warning! The human body should not be exposed to X-rays because of their damaging effect on tissues

---

The crystal in its capillary is attached to a goniometerhead (Figure 1.11). This gadget has two perpendicular arcs, allowing rotation of the

Figure 1.11. A goniometerhead.

crystal along two perpendicular axes. For further adjustment and centering, the upper part of the goniometerhead can be moved along two perpendicular sledges. After preadjustment under a microscope, the goniometerhead is screwed on to an area detector.[1] At this point you need the assistance of a colleague with experience in handling the area detector.

Be sure that the crystal is in the X-ray beam path and adjust it with its 4-fold symmetry axis along the direction of the X-ray beam. The crystal-to-detector distance must be 5–6 cm. Observe the X-ray pattern with the crystal oscillating over a tiny angle (e.g., 0.15°). You will see reflection spots arranged in circles (Figure 1.12a). These circles can be regarded as the intersection of a series of parallel planes with a sphere (Figure 1.13). Try to adjust the orientation of the crystal such that the planes are perpendicular to the X-ray beam. The inner circle will then disappear and the others will be concentric around the beam, which is hidden by the backstop. From this position oscillate the crystal over a 3° range; many more reflections appear on the screen. The inner circle appears again and you observe that the plane to which it corresponds has spots nicely

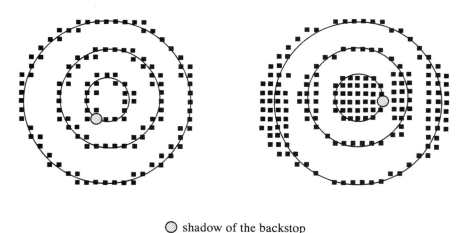

⊚ shadow of the backstop

a   b

Figure 1.12. (a) A schematic representation of the diffraction pattern of a stationary lysozyme crystal. The diffraction spots are arranged in circles. The innermost circle passes through the origin, which is behind the backstop. The latter prevents the strong primary beam to reach the detector. (b) A three degree oscillation picture of the same lysozyme crystal.

---

[1] If an area detector is not available you could use a precession camera or an image plate system.

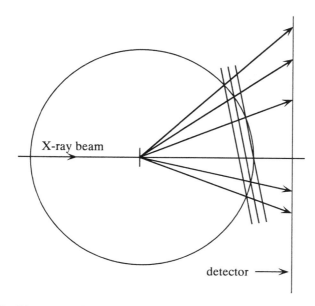

Figure 1.13. Diagrammatic representation of the generation of an X-ray diffraction pattern. The circles shown in Figure 1.12a can be regarded as the
intersection of a series of parallel planes with a sphere. The planes are part of a
lattice composed of diffraction spots. However, diffraction occurs only if the spots
are on, or pass through, the surface of the sphere. The planes are determined by
the lysozyme crystal and the sphere by the wavelength of the X-ray radiation
(Chapter 4).

arranged in perpendicular rows (Figure 1.12b). This is true for all planes
and your crystal produces thousands of diffraction spots in a well-ordered
lattice. From the position of these spots the repeating distances in the
crystal can be derived and from their intensities the structure of the
lysozyme molecules.

The mercury derivative crystals will produce the same pattern as crystals
of the native enzyme. However, close inspection will show that although
the position of the spots is the same, the intensity of the diffracted beams
is slightly different.

1.7. Notes

1. Crystallization: The first essential step in determining the X-ray structure of a protein is to grow crystals of sufficient size and quality. It is
always amazing to see that these big molecules can arrange themselves so
neatly that crystals with flat faces and sharp edges grow. There is great

excitement in the laboratory when the first crystals of a protein appear and the excitement is even greater if those crystals produce a high quality X-ray diffraction pattern. Although more steps are required to complete a structure determination, the growth of good quality crystals indicates a reasonable chance of success and suggests that the protein structure can indeed be determined.

Choosing a crystallization technique and crystallization conditions is largely a matter of taste. The hanging or sitting drop methods are the most popular. Liquid–liquid diffusion in a small-bore capillary has the advantage of being somewhat slower in reaching equilibrium and this is sometimes an advantage. Dialysis is not frequently used but it must be applied for salting-in precipitation.

After preparing the crystallization experiments, leave them in a quiet place with a minimum of vibration and keep the temperature fairly constant.

Protein crystallization is in essence a trial-and-error method, and the results are usually unpredictable. Serendipity also plays a role. In the author's laboratory a failure in the air-conditioning system, causing an unexpected rise in temperature, led to the growth of perfect crystals of an enzyme, which could not be obtained under more normal circumstances.

Commercially available robots make it possible to perform more experiments in the same time and to determine the optimum crystallization conditions more quickly. However, human intelligence is still required to tell the robot what to do.

If crystallization does not occur, even after many experiments, the following can be tried:

a. Crystallization of a homologous protein from another source.
b. Crystallization of one or more proteolytic fragments of the protein. The polypeptide chain of the native protein is split by a proteolytic enzyme at a limited number of positions, or, alternatively, the required fragment is expressed and purified from a bacterial or eucariotic expression system.

2. Mounting the crystals: X-ray capillaries are very fragile because of their thin glass wall. Why not mount them in a stronger glass tube? Although you can see the crystal just as well through a much thicker glass

Table 1.2. Transmission through Pyrex Glass

	Glass thickness (mm)	Transmission (%)
$\lambda = 1.54\ \text{Å}$	0.01	93
(X-ray tube with	0.1	50
Cu-anode)	1.0	0.01
$\lambda = 0.71\ \text{Å}$	0.1	93
(X-ray tube with	1.0	40
Mo-anode)		

wall, X-rays cannot. A glass plate of 0.01 mm thickness has 93% trans-mission for the common X-ray wavelength of 1.54 Å, but this diminishes exponentially with increasing thickness of the glass (Table 1.2). The data show that transmission is highly dependent on the X-ray wavelength and is much higher for a shorter wavelength. This is one of the reasons that X-ray data collection at a synchrotron is preferably done around 1 Å instead of 1.5 Å, the usual wavelength in the home laboratory. The disadvantage of a lower diffraction intensity at shorter wavelength is compensated by the strong intensity of the synchrotron X-ray beam. Other advantages of a shorter wavelength are the more favorable crystal lifetime and better response for a detector with a fluorescent screen.

X-ray wavelength and atomic distances are usually expressed in Ångström units: $1 \text{ Å} = 10^{-10} \text{ m} = 10^{-1} \text{ nm}$. Although the Ångström unit is not an SI unit it has the advantage of giving simpler numbers. For example, the $C-C$ distance in ethane is 1.54 Å, and the most frequently used wavelength in the X-ray diffraction of proteins is 1.5418 Å. Some-times the photon energy (E) is given instead of the wavelength λ. The relationship is λ (in Å) $= 12.398/E$ (in keV).

3. The X-ray pattern: You have observed X-ray diffraction spots nicely arranged in rows. You should compare this with the laser + electron microscope grid experiment. There you saw a two-dimensional diffraction pattern with rows of spots produced by a two-dimensional grid and the diffraction pattern was reciprocal to the actual electron microscope grid. A crystal can be regarded as a three-dimensional grid and you can imagine that this will produce a three-dimensional X-ray diffraction pattern. As with the electron microscope grid, the pattern is reciprocal to the crystal lattice.

The planes that intersect the sphere in Figure 1.13 are layers in a three-dimensional lattice, which is not the crystal lattice, but its reciprocal lattice. The unit distances in this lattice are related reciprocally to the unit distances in the crystal and that is why the lattice is called a reciprocal lattice. Each reciprocal lattice point corresponds to one diffracted beam. The reciprocal lattice is an imaginary but extremely convenient concept for determining the direction of the diffracted beams. If the crystal rotates, the reciprocal lattice rotates with it. In an X-ray diffraction experiment the direction of the diffracted beams depends on two factors: the unit cell distances in the crystal from which the unit cell distances in the reciprocal lattice are derived and the X-ray wavelength.

As indicated in Figure 1.13 diffraction conditions are determined not only by the reciprocal lattice, but also by the radius of the sphere, which is called the sphere of reflection or "Ewald sphere."[2] Its radius is

---

[2] Paul P. Ewald, 1888–1895, professor of physics at several universities in Europe and the United States, was the first to apply the reciprocal lattice and the sphere named after him to the interpretation of an X-ray diffraction pattern.

reciprocal to the wavelength λ; it is equal to $1/\lambda$. From the diffraction experiment with lysozyme you can determine that not all diffracted beams occur at the same time. Only the ones corresponding to reciprocal lattice points on the Ewald sphere in Figure 1.13 are actually observed. Other points can be brought to diffraction by rotating the crystal and with it the reciprocal lattice, to bring these new lattice points on the sphere.

Diffracted beams are often called "reflections." This is due to the fact that each of them can be regarded as a reflection of the primary beam against planes in the crystal. Why this is so will be explained in Chapter 4.

Summary

Protein crystal growth is mainly a trial-and-error process but the primary rule is that the protein should be as pure as possible. The higher the purity, the better the chance of growing crystals by batch crystallization, by liquid–liquid or vapor diffusion, or by dialysis. The experiment with hen egg white lysozyme, the growth of nicely shaped crystals, and the huge number of reflections in the X-ray diffraction pattern have provided an introduction to protein X-ray crystallography. From the X-ray diffraction experiment it became clear that the X-ray pattern can be regarded as derived from the intersection of a lattice and a sphere. This lattice is not the crystal lattice but the reciprocal lattice, which is an imaginary lattice related to the crystal lattice in a reciprocal way. The sphere is called the Ewald sphere and has a radius of $1/\lambda$.

Chapter 2
X-ray Sources and Detectors

2.1. Introduction

In Chapter 1 you learned how crystals of a protein can be grown and you observed a diffraction pattern. The crystalline form of a protein is required to determine the protein's structure by X-ray diffraction, but equally necessary are the tools for recording the diffraction pattern. These will be described in this chapter on hardware. The various X-ray sources and their special properties are discussed, followed by a description of cameras and detectors for quantitative and qualitative X-ray data collection.

2.2. X-ray Sources

The main pieces of hardware needed for the collection of X-ray diffraction data are an X-ray source and an X-ray detector. X-rays are electromagnetic radiation with wavelengths of 10^{-7}–10^{-11} m (1000–0.1 Å). Such radiation was discovered by Roentgen[1] in 1895, but as the nature of the radiation was not yet understood, Roentgen called them X-rays. Von Laue's[2] diffraction theory, which he developed around 1910, inspired his assistants, Friedrich and Knipping, to use a crystal as a diffraction grating. Their results, published in 1912, were direct

---

[1] Wilhelm Conrad Roentgen, 1845–1925, discovered X-rays on November 8, 1895 in Würzburg, Germany.
[2] Max von Laue, 1879–1960, German physicist, developed the theory of X-ray diffraction by a three-dimensional lattice.

proof for the existence of lattices in crystals and for the wave nature of X-rays.

For the X-ray diffraction experiment with lysozyme crystals in Chapter 1 an X-ray generator in the home laboratory was used. These are typically either a generator with a sealed tube and fixed target or a more powerful system with a rotating anode. Both instruments are push-button operated but require some maintenance:

1. a sealed tube must be replaced if the filament burns out and
2. a rotating anode tube occasionally needs a new filament (cathode) and new seals.

Particle accelerators as synchrotrons and storage rings are the most powerful X-ray sources, but they are so complicated technically that the X-ray crystallographer is just a user at the front end. Because protein molecules are very large, their crystals diffract X-ray beams much less than do crystals of small molecules. The reason is that diffraction is a cooperative effect between the molecules in the crystal; for larger molecules there are fewer in a crystal of same size and therefore the diffracted intensity is lower. Moreover, proteins consist mainly of C, N, and O. These are light elements with only a few electrons (6–8) per atom. Since the electrons are responsible for the diffraction, atoms of these light elements scatter X-rays much more weakly than do heavier elements lower in the periodic table. Because of this phenomenon of relatively low scattering power, protein crystallographers prefer a high intensity source: a rotating anode tube rather than a sealed tube. For crystals of a very small size (<0.1 mm) or with extremely large molecules, synchrotron radiation is required for data collection.

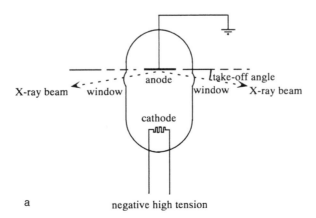

Figure 2.1. (a) A schematic drawing of a sealed X-ray tube. The windows are made of thin berylium foil that has a low X-ray absorption. (b) Cross section of an X-ray tube: Courtesy Philips, Eindhoven, The Netherlands.

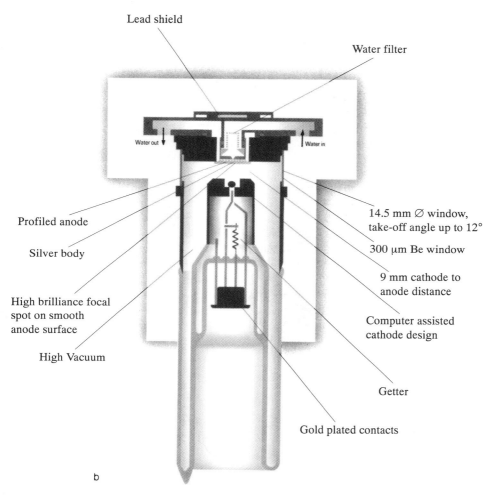

Lead shield

Water filter

Water out

Water in

Profiled anode

Silver body

High brilliance focal
spot on smooth
anode surface

High Vacuum

14.5 mm ⌀ window,
take-off angle up to 12°

300 μm Be window

9 mm cathode to
anode distance

Computer assisted
cathode design

Getter

Gold plated contacts

b

Figure 2.1. *Continued*

   Although X-ray generators with a sealed tube produce a relatively low
beam intensity, they have not disappeared from protein crystallography
laboratories because of their ease of operation and simple maintenance.
For preliminary qualitative work they are fine, especially for relatively
small proteins. Sealed tubes have a guaranteed life time of 1000 hr, but in
practice they last much longer: 1 year of continuous operation is not
unusual. Quantitative measurements should be collected with a rotating
anode tube as X-ray source; they have a radiation intensity approximately
four times higher than that of a sealed X-ray tube. Although your part in
operating an X-ray source mainly consists of pushing the correct buttons,

you should nevertheless be aware of the properties of radiation from various sources. Therefore, we shall discuss them in more detail.

2.2.1. Sealed X-ray Tubes

In a sealed X-ray tube a cathode emits electrons (Figure 2.1). Because the tube is under vacuum and the cathode is at a high negative potential with respect to the metal anode, the electrons are accelerated and reach the anode at high speed. For protein X-ray diffraction the anode is usually a copper plate onto which the electron beam is focused, to a focal spot, normally of 0.4×8 mm. Most of the electron energy is converted to heat, which is removed by cooling the anode, usually with water. However, a small part of the energy is emitted as X-rays in two different ways: as a smooth function of the wavelength and as sharp peaks at specific wavelengths (Figure 2.2). The continuous region is due to the physical phenomenon that decelerated (or accelerated) charged particles emit electromagnetic radiation called "Bremsstrahlung." This region has a sharp cut-off at the short wavelength side. At this edge the X-ray photons obtain their full energy from the electrons when they reach the anode. The electron energy is $e \times$ (accelerating voltage V), where e is the electron charge. The photon energy is $h\nu = h \times (c/\lambda)$, where h is Planck's constant, ν is the frequency of the radiation, c is the speed of light, and λ is the wavelength. Therefore,

$$\lambda_{\min} = \frac{h \times c}{e \times V} = \frac{12.4}{V}$$

where V is in kilovolts. At $V = 40$ kV the cut-off edge is at 0.31 Å.

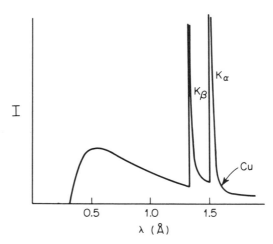

Figure 2.2. The spectrum from an X-ray tube with a copper anode. It shows a continuous spectrum and in addition two sharp peaks due to quantized electrons in the copper. I is the energy of the emitted radiation on an arbitrary scale.

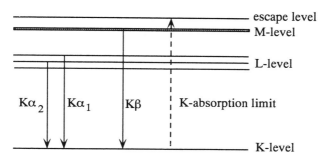

Figure 2.3. Schematic representation of the atomic energy levels and transitions causing characteristic X-ray wavelengths.

The sharp peaks in the spectrum are due to electron transitions between inner orbitals in the atoms of the anode material. The high energy electrons reaching the anode shoot electrons out of low lying orbitals in the anode atoms. Electrons from higher orbitals occupy the empty positions and the energy released in this process is emitted as X-ray radiation of specific wavelength: K_α radiation if it comes from a transition from the L-shell to the K-shell and K_β for a transition from the M- to the K-shell. Because of the fine structure in the L-shell, K_α is split up in $K_{\alpha1}$ and $K_{\alpha2}$. The energy levels in the M-shell are so close that for K_β one wavelength value is given (Figure 2.3).

When copper is the anode material the following values for λ are given:

λ (Å)		
$K_{\alpha1}$	1.54051	The weight average value for $K_{\alpha1}$ and $K_{\alpha2}$ is taken as 1.54178 Å
$K_{\alpha2}$	1.54433	because the intensity of $K_{\alpha1}$ is twice that of $K_{\alpha2}$
K_β	1.39217	

For emission of the characteristic lines in the spectrum a minimum excitation voltage is required. For example, for the emission of the CuK_α line, V should be at least 8 kV. If a higher voltage is applied the intensity of the line is stronger with respect to the continuous radiation, up to about $V/V_{min} = 4$. The intensity of the line is also proportional to the tube current, at least as long as the anode is not overloaded. A normal setting is $V = 40$ kV with a tube current of 37 mA for a 1.5-kW tube.

2.2.2. Rotating Anode Tubes

The heating of the anode caused by the electron beam at the focal spot limits the maximum power of the tube. Too much power would ruin the anode. This limit can be moved to a higher power loading if the anode is a rotating cylinder instead of a fixed piece of metal. With a rotating anode tube small source widths (0.1–0.2 mm) with a high brilliance[3] are possible. The advantage over the sealed tube is the higher radiation intensity, but a disadvantage is that it requires continuous pumping to keep the vacuum at the required level.

The take-off angle (Figure 2.1a) is usually chosen near 4°. Then the observer "sees" the focal spot with dimensions of 0.4×0.5 mm for a sealed tube or smaller for a rotating anode tube. Higher brilliance could be obtained with a smaller take-off angle, but a smaller angle results in a longer X-ray path in the anode material and, therefore, higher absorption and lower beam intensity.

2.2.3. Synchrotron Radiation

Synchrotrons are devices for circulating electrically charged particles (negatively charged electrons or positively charged positrons) at nearly the speed of light [a more detailed discussion is given in Helliwell (1992)]. The particles are injected into the storage ring directly from a linear accelerator or through a booster synchrotron (Figure 2.4a). Originally these machines were designed for use in high energy physics as particle colliders. When the particle beam changes direction the electrons or positrons are accelerated toward the center of the ring and therefore emit electromagnetic radiation, and consequently lose energy. This energy loss is compensated for by a radiofrequency input at each cycle. The physicists' main aim was to study colliding particles; they were not interested in the radiation, which they regarded as an annoying byproduct and wasted energy. However, chemists and molecular biologists discovered (Rosenbaum et al., 1971) that the radiation was a useful and extremely powerful tool for their studies to the extent that radiation-dedicated synchrotrons have been constructed.

Synchrotrons are extremely large and expensive facilities, the ring having a diameter of 10 to a few hundred m. The trajectory of the particles is determined by their energy and by the magnetic field, which causes the charged particles to change their direction. There are four types of magnetic devices in storage rings:

1. *bending magnets* needed to guide the electrons in their orbit and three devices that extend the spectrum to shorter wavelength. These devices can be inserted into straight sections and give no net displacement of the particle trajectory:

2. a *wavelength shifter*, with a stronger local magnetic field and a sharper curvature than the bending magnets,

---

[3] Brilliance is defined as number of photons/sec/mrad^2/mm^2/0.1% relative bandwidth.

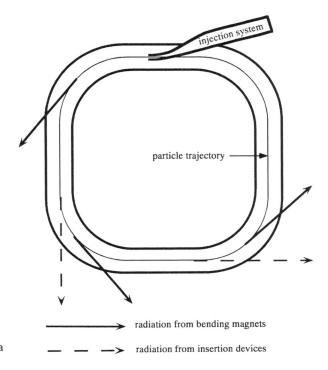

injection system

particle trajectory

→ radiation from bending magnets

a   − − −→ radiation from insertion devices

b

Figure 2.4. (a) A schematic picture of a particle storage ring with an injector for the charged particles. Radiation can be obtained from the particles while passing the bending magnets. In the straight sections of the ring devices can be inserted, e.g., a wiggler or undulator, that produce much higher intensity radiation than the bending magnets. (b) An artist's picture of the European Synchrotron Radiation Facility in Grenoble.

3. a *wiggler*, producing a number of sharp extra bends in the electron trajectory, and
4. an *undulator*, similar to a wiggler but with the difference that interference effects cause the emission of radiation at more specific wavelengths.

Because freely traveling electrons (and positrons) are not quantized, the radiation ranges over a wide wavelength region depending on the energy of the charged particles and on the strength of the magnetic field. A useful quantity is the median of the distribution of power over the spectral region, called the critical photon energy E_c, it divides the power spectrum into two equal parts.

$$E_c = 0.665E^2B \quad \text{or} \quad \lambda_c = \frac{18.64}{E^2B} \qquad (2.1)$$

E_c is in keV and E is the circulating power in GeV:

$$E = \frac{\text{particle energy} \times \text{current}}{\text{revolution frequency}} \qquad (2.2)$$

B is the magnetic field strength in Tesla and λ_c is in Å. The main photon flux is close to E_c, but above E_c it drops exponentially as a function of the photon energy (Figure 2.5). The European Synchrotron Radiation Facility (ESRF) in Grenoble (Figure 2.4b) has a circumference of 844.39 m, is operated with an energy of 6 GeV, and has bending magnets with 0.86 T field strength. Therefore λ_c is 0.6 Å.

The radius of a storage ring depends on E and B: it is proportional to E and inversely proportional to B. If one wants to design a synchrotron

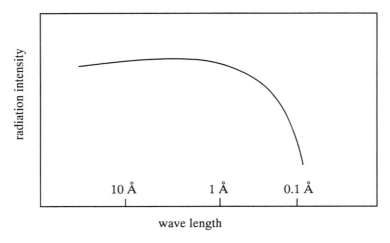

Figure 2.5. A typical curve of radiation intensity as a function of wavelength for a synchrotron radiation source.

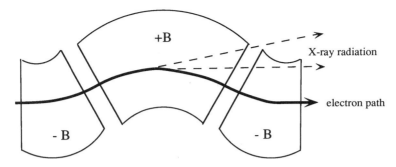

Figure 2.6. A wave length shifter. B is the magnetic field strength. The emission of radiation comes essentially from the top of the bump in the electron path.

for the home laboratory, relatively small dimensions of the instrument are required and therefore E should be low and B high. The most powerful superconducting magnets for laboratory use are found in NMR instruments; they have a field strength of $B = 15\,T$. If λ_c has to be $0.6\,\text{Å}$ for the home instrument, as it is for the Grenoble facility, and assume that suitable magnets of $15\,T$ can be constructed, we calculate for E:

$$E = \sqrt{E^2} = \sqrt{\frac{18.64}{\lambda_c \times B}} = 1.44\,\text{GeV}$$

The diameter of the instrument would then be approximately $4\,m$. However, the feasibility of such a project also depends on other factors, such as the injection system, which must be at least $50\,m$ in length. Other problems are the spreading of the beam and the slow rate at which the magnetic field can be increased after beam injection.

1. The wavelength shifter: According to Eq. (2.1) the critical photon energy E_c is determined by E and B. E is everywhere constant in the ring, but the magnetic field strength B can be increased locally resulting in a decrease of E_c and the production of higher intensity X-rays at shorter wavelength. A schematic picture of a wavelength shifter is shown in Figure 2.6.

2. The multipole wiggler: A series of wavelength shifters constitutes a multipole wiggler (Figure 2.7). The radiation from consecutive magnets is independent and adds up incoherently in the general direction of propagation of the electron beam. The total flux is simply $2N$ times the flux generated by a single period, where N is the number of periods. It is easily tunable to the desired wavelength.

3. The undulator: Undulators are multipole wigglers but with moderate magnetic fields and a large number of poles. The effect of this difference is that in the undulator strong interference occurs between the radiation

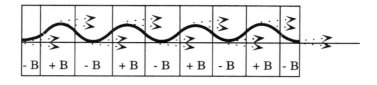

——— electron path

· · · ·> X-ray radiation

Figure 2.7. A multipole wiggler. B is the magnetic field strength.

from the consecutive magnets, which results in a spectral profile with a peak at a specific wavelength and a few harmonics. Therefore this radiation is not tunable, but the advantage for some experiments is that a monochromator is not required. Its emitted intensity can be N^2 times that of a single period, with N the number of poles.

2.2.3.1. Properties of Synchrotron Radiation

Intensity

The main advantage of synchrotron radiation for X-ray diffraction is its high intensity, which is two orders of magnitude stronger than for a conventional X-ray tube, at least for radiation from a bending magnet. For radiation from a multipole wiggler or an undulator it is again a few orders of magnitude stronger. This high intensity is profitably used by protein X-ray crystallographers for data collection on weakly diffracting specimens, such as very tiny crystals or crystals with extremely large unit cells. Another advantage is the low divergence of the beam resulting in sharper diffraction spots.

Tunability

Synchrotron radiation also differs from tube radiation in its tunability. Any suitable wavelength in the spectral range can be selected with a monochromator. This property is used in multiple wavelength anomalous dispersion (Section 9.5) and for Laue diffraction studies (Chapter 12). In the latter method a wide spectral range is used. Several types of modern detectors for X-ray radiation have fluorescent material as their X-ray-sensitive component. This is more sensitive for radiation with lower wavelength, e.g., 1 Å instead of the conventional 1.5 Å from a copper target. Therefore, in protein X-ray diffraction experiments, a synchrotron is tuned to 1 Å or even a shorter wavelength. The additional advantage of this shorter wavelength is lower absorption along its path and in the crystal. Moreover, radiation damage to the protein crystal is appreciably reduced and often all required diffraction data can be collected from just one crystal.

Lifetime

The lifetime of a storage ring filling is limited. Typical lifetimes are between a few and several hours. When the intensity of the radiation has fallen to a certain minimum value, a new injection is required. The factors determining the lifetime are many and of a complicated nature, e.g., collision with residual gas atoms. The positively charged ions that form accumulate in the electron beam, leading to instabilities in the beam due to the continuous loss of particles. A beam composed of positrons has a longer lifetime than an electron beam, because positrons repel positive ions created in the residual gas, thus avoiding ion trapping in the beam. The beam lifetime in the European Synchrotron Radiation Facility is expected to be approximately 10 hr.

Time Structure

Synchrotron radiation, in contrast to X-ray tube radiation, is produced in flashes by the circulating bunches of charged particles. The ESRF will operate in single bunch or multibunch mode with a bunch length of 16–50 psec. In principle this property could be used for time-resolved measurements in the microsecond range.

Polarization

The X-ray beam from an X-ray tube is not polarized; synchrotron radiation is highly polarized. If the radiation from a bending magnet is observed in the plane of the orbit, it is fully polarized with the electric vector in the orbit plane (parallel polarization). If the observer moves away from the plane, a small perpendicular component is added. The polarization of the X-ray beam from a synchrotron has not yet found extensive application in X-ray diffraction. However, it must be considered when applying the correct polarization factor (Section 4.15.1). Moreover the polarization of the beam has an effect on the so-called anomalous X-ray scattering of atoms, which occurs when the X-ray wavelength approaches an absorption edge wavelength (Section 7.8).

2.3. Monochromators

Except for Laue diffraction, where the crystal is exposed to a spectrum of wavelengths (Chapter 12), monochromatic X-rays are used in all other diffraction methods. Therefore, a narrow wavelength band must be

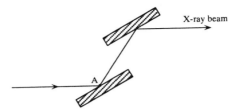

Figure 2.8. A double monochromator. The rotation axis for changing the wave-length is at A, perpendicular to the plane of the page. This gives a small vertical shift to the exit beam after rotation, but the incoming and exit beam are parallel.

selected from the spectrum supplied by the source. If a sealed X-ray tube with copper anode is the radiation source, the wavelength selected for protein X-ray diffraction is the high intensity K_α doublet ($\lambda = 1.5418\,\text{Å}$). The K_β radiation can be removed with a nickel filter. A nickel foil of 0.013 mm thickness reduces the K_β radiation to 2% and K_α to 66% of the original intensity. The continuous spectrum is also reduced but certainly not eliminated.

Much cleaner radiation can be obtained with a monochromator. For X-ray radiation from a tube the monochromator is a piece of graphite that reflects X-rays of 1.5418 Å against its layer structure at a scanning angle of 13.1°. For synchrotron radiation the preferable monochromators are made of germanium or silicium because they select a wavelength band two orders of magnitude narrower ($\delta\lambda/\lambda = 10^{-4}-10^{-5}$). Monochromators for synchrotron radiation are of the single or double type. Single type monochromators can be either flat or bent. The advantage of the bent monochromators is that they focus the divergent beam from the syn-chrotron, preferably onto the specimen. The focusing is in one direction only, producing a line focus. Focusing in the other direction is obtained with toroidal quartz mirrors, which are sometimes gold or platinum coated.

The single type monochromators have a disadvantage: if they are tuned to another wavelength, the scanning angle of the monochromator changes and the entire X-ray diffraction equipment must be moved. This is not necessary for a double monochromator, where the direction of the beam is independent of the wavelength (Figure 2.8).

2.4. Introduction to Cameras and Detectors

In the X-ray diffraction experiment with lysozyme in Chapter 1 a dif-fraction pattern was observed that could be regarded as corresponding to a three-dimensional lattice, reciprocal to the actual crystal lattice. For a crystal structure determination the intensities of all (or a great many) diffracted beams must be measured. To do so, all corresponding reciprocal lattice points must be brought to diffracting condition by rotating the lattice (that is, by rotating the crystal) until the required reciprocal lattice points are on the sphere with radius $1/\lambda$. It follows that an X-ray diffraction instrument consists of two main parts:

1. A mechanical part for rotating the crystal.
2. A detecting device to measure the position and the intensity of the diffracted beams; it should be noted that this intensity is always the result of two measurements: the total intensity in the direction of the diffracted beam and, subtracted from it, the background scattering in that same direction.[4]

In the last decade data collection in protein crystallography has changed tremendously. Formerly a well-equipped protein crystallography laboratory had some precession cameras, a rotation camera, and a diffractometer. In both types of camera, X-ray film was the detector. The precession camera has the advantage of giving an undistorted image of the reciprocal lattice. Unit cell dimensions and symmetry in the crystal can easily be derived from such an undistorted image as can the quality of the crystal. For full three-dimensional X-ray data collection the precession camera is not suitable, because it registers one reciprocal layer per exposure.

A rotation camera registers the data more efficiently, but the recognition of the diffraction spots is more complicated. Moreover, for each exposure the crystal is oscillated over a small angle (e.g., 2°) to avoid overlap of spots. Depending on the symmetry in the crystal the total oscillating range for a complete data collection may be 60°, 90°, or 180°. As a result dozens of filmpacks must be exposed, usually with two or three films in a pack. It is not uncommon to have a pile of 100–300 films to be digitized by a densitometer and processed further with suitable software. This is time consuming and is done only if a small pixel size ($<100\,\mu$m) is required, as for the measurement of small diffraction spots obtained with synchrotron radiation from a virus crystal.

The third classical detector is a computer-controlled diffractometer, which has a single counter, normally a scintillation counter. This has been the workhorse in protein X-ray crystallography and still is so for the X-ray diffraction of small molecules. It measures the diffracted beams with high accuracy but only one at a time. Although it is computer-controlled and automatically finds the diffracted beams, it is extremely slow.

The classical picture has been changed completely by the introduction of much faster image plates and electronic area detectors. Rotation

---

[4] Background scattering is mainly caused by the air through which the X-ray beam passes from the collimator to the backstop. If the airpath is long and absorption serious, it can appreciably be reduced if a cone filled with helium is put between the crystal and the plate.

"cameras" are now equipped with such a detector, either image plate or electronic area detector, though the principle of the "camera" has not changed. With an image plate, which must be read after each exposure, the data collection is still in the "film mode," that is contiguous oscillations over a small angle and reading (analogous to film processing) after each exposure. If the instrument has an electronic area detector the oscillations are much smaller (e.g., 0.1°), and the data are immediately processed by an on-line data acquisition system.

Before describing these instruments in some detail, we shall discuss the various types of X-ray-detecting systems including their advantages and disadvantages.

2.5. Detectors

2.5.1. Single Photon Counters

Single photon counters have been used since the early days of X-ray diffraction and are now usually of the scintillation type. They give very accurate results, but because they measure X-ray reflections sequentially it takes several weeks to collect a complete data set from a protein crystal (of the order of 10,000 to 100,000 reflections). If time is not a limiting factor, or if the protein is not too large and fairly stable in the X-ray beam, the combination of a scintillation counter and a computer-controlled diffractometer is a useful instrument in a protein crystallography laboratory. However, it is being replaced by the much faster image plate and electronic area detector systems.

Although the instrument with a single photon counter is no longer used very often for data collection in protein X-ray crystallography, we shall present its mechanical construction, because the angle nomenclature is standard in X-ray crystallography (Figure 2.9a). In this diffractometer the X-ray beam, the counter, and the crystal are in a horizontal plane. To measure the intensity of a diffracted beam the crystal must be oriented such that this beam will also be in the horizontal plane. This orientation is achieved by the rotation of the crystal around three axes: the ϕ-, the ω-, and the χ-axis (see Figure 2.9a). The counter can be rotated in the horizontal plane around the 2θ-axis, which is coincident with, but independent of the ω-axis. Data collection is done either with the ω- and the 2θ-axis coupled or with the 2θ-axis fixed and the crystal scanned by rotation around the ω-axis.

An alternative construction is found in the CAD4 diffractometer produced by Delft Instruments. It has two advantages over the classical design. The rather bulky χ circle is absent; instead the instrument has another axis, the oblique κ-axis (Figure 2.9b). The χ rotation is mimicked by a combined rotation around the κ-, ϕ-, and ω-axes. The other ad-

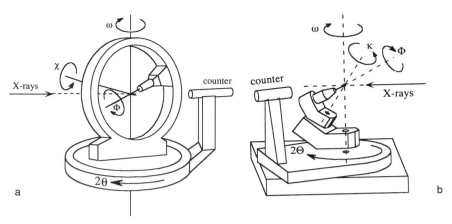

Figure 2.9. (a) In this four circle diffractometer of classical design, the crystal is located in the center of the large circle. It can be rotated around three axes: by φ around the axis of the goniometer head, by ω around the vertical axis, and by χ through sliding of the block that holds the goniometerhead, along the large circle. The counter can be rotated around a fourth axis by the angle 2θ; this axis is coincident with the ω-axis. (b) A four circle diffractometer with the κ construction. The κ-axis is at 50° with respect to the ω-axis. The φ rotation is again around the goniometerhead axis. The ω- and the 2θ-axes are as in the classical design. The crystal is at the intersection of all axes.

vantage is that rotation around an axis is mechanically more accurate than sliding along an arc.

2.5.2. Photographic Film

Photographic film is a classical detector for X-ray radiation, but it is not used much anymore because of the availability of far more sensitive image plates and area detectors. For historical reasons we shall devote one paragraph to this classical detector.

X-ray film is double-coated photographic film. The double coating of the film base prevents curling and provides a thick layer of sensitive silver-base material, which absorbs approximately two-thirds of the incoming photons. Processing of X-ray film is—although not difficult—somewhat cumbersome and time consuming because of the developing process, the labeling of the films, as well as the handling of chemicals. Furthermore, for quantitative work, the density of the spots on the film must be measured with a densitometer. The density is defined as $\log(I_0/I)$, where I_0 is the intensity of the light beam in the densitometer before, and I after it has passed the film. The density of the spots is (up to a certain level) proportional to the number of X-ray photons that are absorbed by the film. Density, plotted as a function of the exposure time, is a straight line over a long exposure range but levels off at very high density values. These higher density values should be used only with careful calibration. This limited dynamic range (1:200) requires that for quantitative measurements, in which the full range of X-ray intensities must be measured, a pack of three consecutive films is used. The weakest spots are measured on the first film and the stronger ones on the last film. Altogether, photographic film is a simple but rather slow and in the long run even expensive detector.

2.5.3. Image Plates

Image plates are used in the same manner as X-ray film but have several advantages. They are therefore replacing conventional X-ray film in most laboratories. Image plates are made by depositing a thin layer of an inorganic storage phosphor on a flat base. X-ray photons excite electrons in the material to higher energy levels. Part of this energy is emitted very soon as normal fluorescent light in the visible wavelength region. However, an appreciable amount of energy is retained in the material by electrons trapped in color centers; it is dissipated only slowly over a period of several days. This stored energy is released on illumination with light. In practical applications a red laser is used for scanning the plate and blue light is emitted. The red light is filtered away and the blue light is measured with a photomultiplier (Figure 2.10a). With certain precautions the light emitted is proportional to the number of photons to which that particular position of the plate was exposed. The pixel size depends mainly on the reading system and is between 100×100 and $200 \times 200\,\mu m^2$.

Image plates are at least 10 times more sensitive than X-ray film and their dynamic range is much wider $(1:10^4 - 10^5)$. The entire range from strong to weak reflections can therefore be collected with one exposure on a single plate. The plates can be erased by exposure to intense white light and used repeatedly. Another advantage for application with synchrotron radiation is their high sensitivity at shorter wavelengths (e.g., $0.65\,\text{Å}$). A further advantage of short wavelengths is that the absorption of the X-ray beam in the protein crystal becomes negligible and no absorption corrections are required. But image plates are similar to photographic film in the sense that they require a multistep process: exposure as the first step and processing (in this case reading) as the second step. Reading takes only a few minutes.

The first commercially available instruments had a rather small size plate but in new models this has been increased (from 18 to 30 cm for the Mar Research instrument Figure 2.10b; the Rigaku R-AXIS II system has rectangular plates of $20 \times 20\,\text{cm}$). If synchrotron radiation is used, the

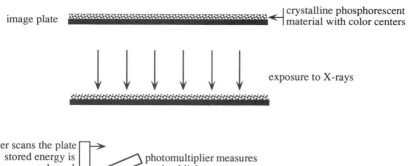

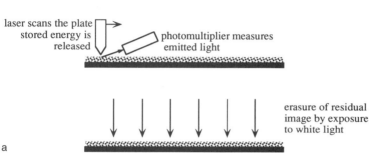

a

b

Figure 2.10. (a) The principle of the image plate as an X-ray detector. (b) The image plate instrument produced by Mar Research, Hamburg Germany. It has a circular plate; for scanning the plate rotates and the laser scans along a radial line. Courtesy Mar Research.

smaller size is not a problem because with the shorter wavelength the diffraction pattern is more compressed. A disadvantage of image plates is that the stored image fades away gradually. This decay is rather rapid in the first few minutes but then slows down: it takes a few hundred hours for a 50% decrease in the stored energy.

2.5.4. Area Detectors

Although photographic film and image plates are area detectors, the use of this term is restricted to electronic devices that detect X-ray photons on a two-dimensional surface and process the signal immediately after photon detection. They are also called position sensitive detectors, because both the intensity of a diffracted beam and the position where it hits the detector are determined. A basic difference with image plate and photographic film is that area detectors scan through a diffraction spot every 0.1° or so, giving a three-dimensional picture of the spot. In contrast, for film and image plates a much larger oscillation angle (e.g., 2°) is used for each exposure and therefore no profile is obtained of the diffraction spot in the oscillation direction.

Area detectors are currently based on either a gas-filled ionization chamber or an image intensifier coupled to a video system. The gas-filled chambers are essentially single photon counting X-ray detectors. Their construction is shown schematically in Figure 2.11a.

In the absorption gap X-ray photons cause ionization of gas atoms with the formation of ions and electrons. The liberated electrons ionize neighboring gas atoms by collision with the result that about 300 ions and electron pairs are formed by the absorption of a single 8-keV (λ = 1.55 Å) X-ray photon. This is not enough for a measurable signal and, therefore, amplification is applied by accelerating the electrons in an electric field between a cathode and an anode. The anode consists of many parallel wires, 1–2 mm apart. The accelerated electrons cause secondary ionization, mainly at the anode wires. The electrons hit the anode and the ions the second cathode, which—like the anode—consists of many parallel wires that are perpendicular to the anode wires. The event is now registered as one count and the position of the incident photon is electronically determined. A disadvantage of this type of detector is the limitation in counting rate due to the build-up of charges in the chamber and to limitations in the processing electronics. The peak counting rate is about 10^5 Hz. This prevents the use of area detectors of the multiple wire proportional counter type in X-ray crystallography with high-intensity synchrotron radiation. Another disadvantage of this type of detector, in combination with synchrotron radiation, is the lower sensitivity at shorter wavelength due to poor absorption of the X-ray photons in the gas chamber.

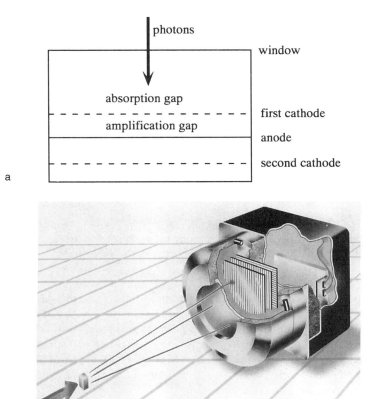

a

b

Figure 2.11. (a) The principle of the design of a multiple wire proportional counter. The gas-filled chamber has two cathodes and an anode, each consisting of parallel wires, either horizontal or vertical. (b) The Siemens area detector, which is filled with xenon gas for maximum absorption of the X-ray beam.

In video-based area detectors the diffraction pattern is collected on a fluorescent screen. The resultant light is amplified with an image intensifier, stored in the target of a video camera tube, read out, and fed into a computer. The system is schematically represented in Figure 2.12. Because of the storage in the video tube the diffraction pattern is integrated over short time periods. Overloading is a less serious problem than for the gas-filled detectors. Therefore, they are better suited for synchrotron radiation than are the gas-filled multiple wire detectors. They have a spatial resolution of about 0.1 mm. Electronic noise is rather high but can be kept under control, e.g., by maintaining the components at a constant temperature. This noise level limits the dynamic range to $1:10^3$. By remeasuring the strong intensities with a lower high voltage setting of the video-camera the dynamic range is increased by a factor of 10.

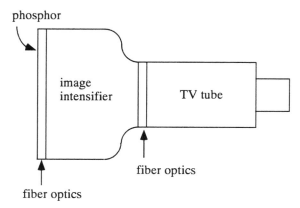

Figure 2.12. An area detector of the television type. X-ray photons enter on the left and are transformed into light photons by the phosphor screen. The light is amplified in the image intensifier and registered in the video tube.

The performance of the two types of area detectors, the gas-filled and the TV systems, is approximately the same. They are both very sensitive, compared with X-ray film. This, and the rapid processing of the data, allows a data collection speed that is about 50 times the speed obtained with X-ray film. Calibration of area detectors is required because of nonuniform sensitivity over the area and geometric distortion of the diffraction pattern in the system.

In another kind of area detector the video tube is replaced by a charge coupled device (CCD), but an optimal design for X-ray work is not yet commercially available. However, much effort is being invested to improve them and to make CCDs of the relatively large size and high sensitivity required for X-ray detection. They have a high dynamic range, combined with excellent spatial resolution, low noise, and high maximum count rate.

2.6. The Precession Camera

You have learned that the diffraction pattern of a crystal can be regarded as originating from a lattice, reciprocal to the crystal lattice. Those reciprocal lattice points that pass through the Ewald sphere (radius $1/\lambda$) are recorded on the detector. In the precession camera the detector is a flat photographic film and the construction of the camera is such that it produces an undistorted image of reciprocal lattice planes. This is a great advantage. Another advantage of the precession camera is that from its pictures the symmetry in the diffraction pattern can be easily recognized. It is no wonder that this camera has found widespread application in protein X-ray crystallography. However, because photographic film is no longer very popular, its significance has been decreased appreciably. Moreover, the precession camera is not suitable for efficient three-dimensional data collection. But if there is still a precession camera in the laboratory, you should record a few reciprocal lattice planes; this will increase your familiarity with the reciprocal lattice concept.

To obtain an undistorted image of a reciprocal lattice plane, the film must always stay parallel to that plane. In Figure 2.13 this is illustrated for a plane through the origin, a zero layer. If crystal and film do not move, this plane is recorded as a diffraction circle on the film corresponding with the intersection of the lattice plane and the Ewald sphere. However, if the normal to the film and the normal to the reciprocal lattice plane are rotated around the direction of the X-ray beam, this precession motion causes the circle of intersection to rotate in the reciprocal lattice plane. As a result, a circular part of the lattice plane is recorded on the film (Figure 2.14). This circular part has its origin in the direct beam and has a radius equal to the diameter of the intersecting circle. Figure 2.15 shows the construction principle of the precession camera as well as a picture of the Delft Instrument camera. If no precautions are taken, spots from other reciprocal lattice planes will also be recorded. This is prevented

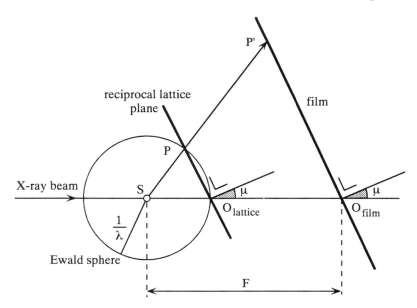

Figure 2.13. A precession camera recording of a plane through the origin of the reciprocal lattice; such a plane is called a zerolayer. μ is the precession angle. The crystal-to-film distance is F. The normal to the film plane rotates around the direction of the X-ray beam in phase with the rotation of the normal to the reciprocal lattice plane.

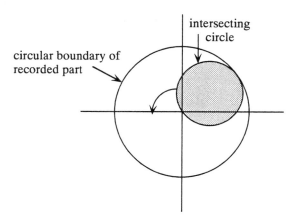

Figure 2.14. The part of the zero layer that is swept out in the precession motion is indicated by the large circle. It is filled by the rotation of the small circle, which is the intersection of the reciprocal lattice plane with the Ewald sphere.

by means of a flat screen that has an annular aperture, as shown in Figure 2.16.

For recording a zero layer with precession angle μ, $r_s/s = \tan\mu$, where r_s is the radius of the annular aperture in the screen and s the distance from the screen to the origin of the diffracted beam, which is at the crystal.

The settings of film and screen for recording nonzero layers are somewhat more complicated. This is illustrated in Figure 2.17 for the film and in Figure 2.18 for the screen. If the image of the nonzero reciprocal lattice layer on the film should be an undistorted one and have the same enlargement factor as for the zero layer, the film must be displaced away from the center of precession in correspondence with the distance d^* of the nonzero reciprocal lattice plane from the origin of the reciprocal lattice. The enlargement factor for a zero layer is equal to

$$\frac{S - O_{\text{film}}}{S - O_{\text{lattice}}} = \frac{F}{1/\lambda} = \lambda F$$

where F is the distance from crystal to film. Therefore, the film displacement D^* is $D^* = d^*\lambda F$. The screen setting can be derived from Figure 2.18:

$$\frac{r_s}{s} = \tan\nu \quad \text{and} \quad d^* = \frac{1}{\lambda}(\cos\mu - \cos\nu); \quad \cos\nu = \cos\mu - \lambda d^*$$

$$\frac{r_s}{s} = \tan\cos^{-1}(\cos\mu - \lambda d^*)^5$$

---

[5]Note that $\cos^{-1}(\cos\mu - \lambda d^*)$ is an inverse function; it is the angle of which the cosine is $(\cos\mu - \lambda d^*)$.

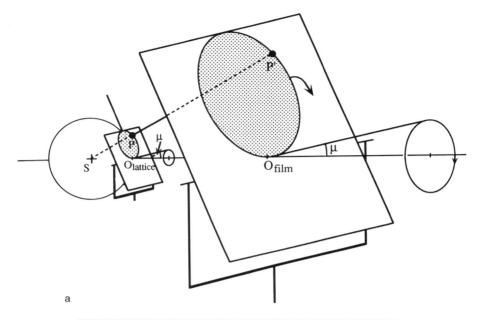

a

b

Figure 2.15. (a) The construction principle of a precession camera. It has two universal joints, the front one for holding the film and the other one for the crystal. The two joints are linked so that film and crystal move together and in phase with a precession angle μ. S is the center of the Ewald sphere. The reciprocal lattice point P is recorded as spot P' on the film. (b) A precession camera as produced by Delft Instruments, Delft, The Netherlands. Courtesy Delft Instruments.

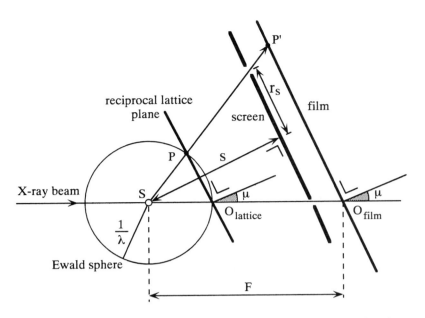

Figure 2.16. The screen allows the zero layer to be recorded on the film but not the other reciprocal lattice planes. r_s is the radius of the annular aperture and s is the distance from crystal to screen. Note that the screen moves with the crystal.

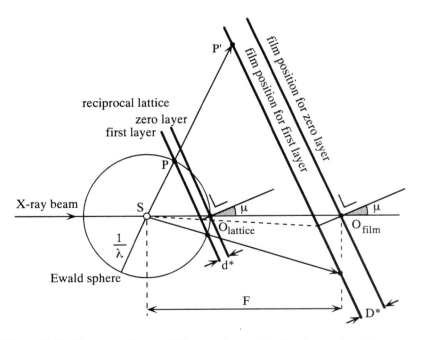

Figure 2.17. For recording a higher reciprocal lattice layer the film must be moved toward the crystal. This is drawn here for a first layer. D^* is proportional to d^*.

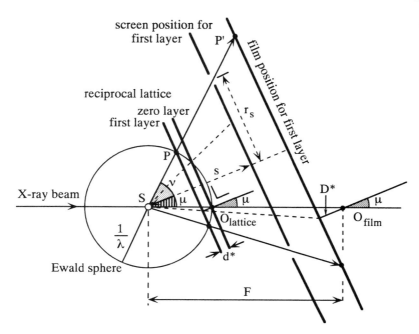

Figure 2.18. The screen position for recording a first reciprocal lattice layer. d^* is equal to the difference between $(1/\lambda)\cos\mu$ and $(1/\lambda)\cos\nu$; $r_s/s = \tan\nu = \tan\cos^{-1}(\cos\mu - \lambda d^*)$.

In practice the required screen is selected from a set of screens with a range of annular apertures provided with the camera. Screens are available with the same radius but different widths for the aperture, such as 1, 2, and 3 mm. If the width chosen is too small only part of the diffracted beams will pass the screen, and if it is chosen much wider than necessary, the background will be stronger and beams from adjacent layers may appear.

If d^* is not yet known, it can be calculated from a "still" picture (no precession motion) with zero precession angle and the d^* direction perpendicular to the film (Figure 2.19)

$$\tan\kappa = \frac{R}{F}; \quad \cos\kappa = \frac{1/\lambda - d^*}{1/\lambda}; \quad d^* = 1/\lambda - 1/\lambda \, \cos\kappa;$$

$$d^* = 1/\lambda - 1/\lambda \, \cos\tan^{-1}\frac{R}{F}; \quad \left(\tan^{-1}\frac{R}{F} = \arctan\frac{R}{F} \text{ is an inverse function}\right)$$

A disadvantage of the precession camera in recording nonzero layers is that they do not show the diffraction spots in the center of the plane

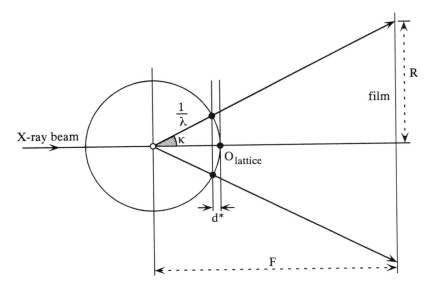

Figure 2.19. The distance d^* can be derived from its relation with the angle κ, which is known from the radius R on the film and the crystal-to-film distance F; d^* is the difference between $1/\lambda$ and $(1/\lambda)\cos\kappa$.

("blind region"). The reason is that these lattice points always stay within the Ewald sphere and never pass through its surface. However, an enormous advantage is the straightforward indexing of the spots and the fact that the symmetry in the diffraction pattern can be easily inspected. There is one point still to be discussed: the orientation of the crystal before a reciprocal lattice plane can be recorded. At zero precession angle ($\mu = 0$) the plane should be perpendicular to the X-ray beam. This is accomplished as follows:

1. Adjust the crystal as accurately as possible under the microscope.
2. Take a "still" picture with $\mu = 0$ and unfiltered radiation (or without a monochromator) and without a screen. An exposure time of a few minutes is sufficient.
3. From the center of the zero layer circle derive the correction angles and apply them to the camera spindle and the goniometer head.
4. Repeat step 3 until the zero layer circle has disappeared.
5. Take a screenless $\mu = 3°$ picture. The zero layer is visible as a circle filled with spots and streaks from the unfiltered radiation. It should have the direct beam position at the circle center. If this is not true adjust again and take a new picture until the correct setting has been reached.
6. Final check. Mount the screen and the radiation filter or monochromator. Set μ at the required value and take a "still." The aperture in

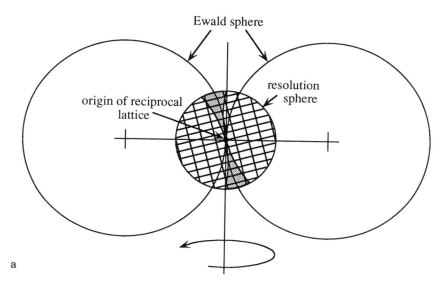

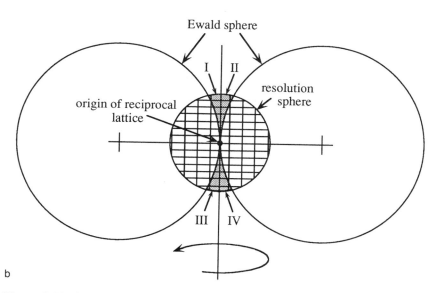

Figure 2.20. (a) For simplicity the Ewald sphere is rotated instead of the re-ciprocal lattice. The vertical line in the center is the rotation axis. In its rotation around the axis the Ewald sphere never passes through the reciprocal lattice regions I–IV. This is the "blind region." (b) Here the reciprocal lattice is in a skew position. In this example the reciprocal lattice has three perpendicular axes and, contrary to the situation in (a), all reflections within one octant formed by the three reciprocal lattice axes can be measured.

the screen causes a circular shadow on the film and in this shadow diffraction spots should be visible.
7. Start the precession motion.

2.7. The Rotation (Oscillation) Instrument

In protein X-ray crystallography efficient data collection is always screen-less, whether photographic film, image plate, or electronic area detector is used. The crystal (and the reciprocal lattice) is rotated in small oscillation steps through the Ewald sphere. The mechanical part of the rotation instrument is very simple: just one rotation axis perpendicular to the direction of the X-ray beam. The detector, film, or fluorescent plate is flat. The consequence of this simple design is that reciprocal lattice points near the rotation axis never pass through the Ewald sphere (Figure 2.20a). Therefore, it requires—in principle—mounting of another crystal in a different orientation to measure the diffracted beams in the blind region. However, this is not always necessary. Suppose the crystal has three perpendicular axes and it is oriented with one of the crystal axes along the rotation axis; then the blind region is indeed a problem: assuming sufficient symmetry in the crystal, the four shaded areas in Figure 2.20a contain reciprocal lattice points belonging to reflections with the same intensity in each region and none of the four regions is recorded. However, if the crystal is slightly misset, the problem is solved because now all reflections in at least one of the four regions can be recorded (Figure 2.20b).

Instruments with an electronic area detector, like the Siemens X-100 (Figure 2.21) and the Fast area detector made by Delft Instruments (Figure 2.22), have a more sophisticated system for rotating the crystal: the Siemens instrument has the classical χ circle and the Delft instrument the κ construction. They allow the crystal to rotate around three axes. Therefore, the same crystal can easily be reoriented with respect to the X-ray beam. The disadvantage of screenless data collection is the seemingly disordered arrangement of the spots on the film or plate. The spot positions are determined by

1. the crystal orientation,
2. the unit cell parameters in the crystal,
3. the crystal-to-film distance and the wavelength, and
4. the film center.

However, this problem can be solved with intelligent software, which can recognize the spots, apply correction factors, and supply the crystallographer with a final data set.

Because of basic differences between instruments equipped with an

Figure 2.21. The Siemens X100 area detector. This instrument is equipped with a multiple wire proportional counter. The crystal can be oriented around three axes. The counter can swing out around another axis for collecting reflections with large diffraction angle. Courtesy Siemens AG, Analytical Systems, Karlsruhe, Germany.

image plate and ones equipped with an electronic area detector, we shall treat them separately.

2.7.1. Rotation Instruments with an Image Plate

This is the modern version of the rotation camera with film, originally designed by U.W. Arndt.[6] Except for the detector, there are no differences in principle between the old film and the new image plate instruments. Data are collected in contiguous oscillation ranges, each of approximately 2°. The exact value is determined by the distance between the reciprocal lattice points, the maximum angle of reflection (resolution), and the width of the spots. In performing the oscillation, the maximum displacement is found for reciprocal lattice points in the plane through the origin perpendicular to the rotation axis at the edge of maximum dif-

---

[6] This instrument is discussed in detail in Arndt and Wonacott (1977).

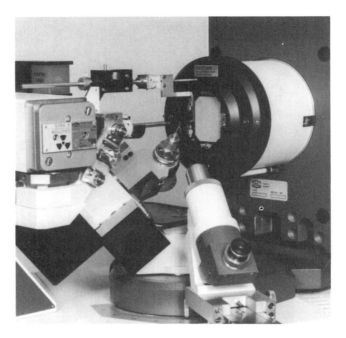

Figure 2.22. The Fast area detector produced by Delft Instruments, Delft, The Netherlands. This is a video type instrument. It has the κ construction for the crystal orientation, derived from the successful CAD-4 four circle diffractometer of this company. The counter can swing out around the 2θ-axis. Courtesy Delft Instruments.

fraction angle (the resolution limit) (Figure 2.23). If the distance between adjacent points at this edge is d^* and their distance from the origin d_{lim}, the angle Δ_1 between them is $\Delta_1 = d^*/d_{lim}$. However, the reflections have a certain angular width Δ_2, determined by the crystal size, its mosaicity, and the divergence of the X-ray beam. Therefore, the maximum oscillation angle is $\Delta_1 - \Delta_2$, because a reflection spot is allowed to move for the next exposure until it reaches the region of its neighbor in the current exposure.

To minimize the number of exposures the oscillation range should be as large as possible and therefore the crystal should be oriented with its shortest reciprocal distance along the rotation axis. (This is not required for area detectors.) Because of the reciprocity between the crystal lattice and the reciprocal lattice, this orientation corresponds with the longest unit distance in the crystal lattice along the rotation axis. Another feature to consider for minimizing the number of exposures is the symmetry in the crystal. This symmetry is also present in the reciprocal lattice. Lattice points related by this symmetry belong to diffracted beams with the same intensity and, therefore, fewer diffracted beams can be measured.

Two disadvantages of the screenless rotation method with image plates are immediately apparent: (1) the background is relatively high and (2) some spots appear partly on one and partly on the next or previous exposure. This second problem is especially severe for large unit cells in the crystal because of the close spacing of reciprocal lattice points and a very small oscillation range. For the solution to this problem the two

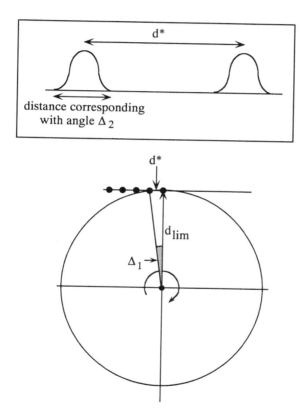

Figure 2.23. In the lower picture the rotation axis is perpendicular to the plane of the paper. The black dots are reciprocal lattice points at the edge of the limiting resolution sphere, which has a radius d_{lim}. The angular distance between adjacent reflections is $\Delta_1 = d^*/d_{lim}$. In the box two adjacent reflections are drawn. Suppose the right reflection has just passed through the Ewald sphere and has caused a spot on the image plate. In the ongoing oscillation the left reflection should not be allowed to reach the Ewald sphere before the right reflection has passed. If it does, then it would overlap the image of the right reflection. Therefore, the maximum oscillation angle permitted is $\Delta_1 - \Delta_2$.

"partials" are treated as individual reflections and their intensities are added in a later stage. This requires extreme precision in the mechanical part of the instrument. Nevertheless, the accuracy in the intensity of reflections composed of partials is somewhat lower than for fully recorded reflections and for this reason they are sometimes completely neglected. The first problem is a signal-to-noise ratio problem. The background noise is proportional to the exposure time and thus to the oscillation angle. The intensity of a fully recorded reflection is independent of the oscillation angle. The signal-to-noise ratio is more favourable if the X-ray beam has a smaller divergence and the mosaic spread of the crystal is low. A smaller oscillation angle also improves the ratio but causes many more reflections to be recorded as partials.

For processing of the data it is an advantage if the crystal is mounted on the instrument in a known orientation, preferably with its longest unit cell dimension along the rotation axis. After exposing the crystal in a few oscillation ranges, the software is able to find an approximate orientation of the crystal with rough unit cell dimensions (if these are not already known). It is assumed that the crystal-to-plate distance and the position of the center of the plate are fixed instrument parameters. The approximate orientation and cell parameters can then be refined during the processing of all the data. In this process a rectangular box is defined around each spot. The integrated intensity in this box is the spot intensity; the background is estimated in the surrounding region. The measurement of weak reflections can be appreciably improved by profile fitting. A reliable two-dimensional spot profile is derived from the strong spots and then applied to the weak ones, of course assuming that the profiles are equal for strong and weak spots.

2.8. Electronic Area Detectors

It was pointed out before that basic differences exist between rotation instruments equipped with an image plate and electronic area detectors. The main difference is that the oscillation angle for the area detectors is much smaller; frames with an oscillation angle of 0.1° are contiguously measured. This is possible due to the immediate processing of the data from each frame. The advantages are that the background is low and that a three-dimensional profile of the reflection spots can be constructed, which is even more favorable for measuring the weak intensities than the two-dimensional profile.

With the more sophisticated mechanical part of the instrument a crystal can be easily adjusted to nearly every orientation. The search for the initial orientation is also easier with these instruments than with a rotation instrument because the smaller oscillation angle defines more precisely the position of a spot.

Radiation Protection

The high energy photons of X-rays have a harmful effect on living tissue. Therefore, they must be used with great care, taking all necessary precautions. Local regulations for the protection of personnel should be obeyed and unauthorized use of X-ray equipment must be forbidden. A somewhat confusing number of radiation units are in use and, therefore, a definition of them will be given.

The Curie (Ci) is the old unit for radioactivity:

$$1\,Ci = 37 \times 10^9 \text{ disintegrations/sec}$$

1 Ci is approximately equal to the activity of 1 g of radium. The Ci is not an SI unit and has therefore been replaced by the Becquerel (Bq), which stands for 1 disintegration/sec:

$$1\,Ci = 37 \times 10^9\,Bq$$

The radiation absorbed dose (rad) is a measure for the amount of radiation that corresponds to the energy absorption in a certain medium, such as a tissue.

1 rad is the dose of radiation corresponding to an energy absorption of 0.01 J/kg medium.

The rad has been replaced by an SI unit: the Gray (Gy), corresponding to 1 J/kg or $1\,m^2/sec^2$.

$$1\,Gy = 100\,rad$$

The relative biological effect (RBE) was introduced because it has been found that the same absorbed dose from different types of radiation does not always have the same harmful effect in biological systems. Therefore, a quality factor Q has been introduced. In the rem (X-ray equivalent man) this quality factor has been taken into account:

$$1\,rem = 1\,rad \times Q$$

The rem has now been replaced by the Sievert (Sv):

$$1\,Sv = 100\,rem$$

A summary of the names and symbols is given in Table 2.1.

Summary

In this chapter X-ray sources were discussed as well as instruments for the registration and measurement of the diffracted beams. Conventional X-ray sources in the laboratory are sealed tubes and tubes with a rotating

Table 2.1. Radiation Units and Symbols

Physical or biological property			SI unit	
Name	Symbol	Value	Symbol	Value
Radioactivity	A	Becquerel	Bq	1 disintegration/sec
		Old: Curie	Ci	$1\,\mathrm{Ci} = 37 \times 10^9\,\mathrm{Bq}$
Absorbed dose	D	Gray	Gy	$1\,\mathrm{J\,kg^{-1}} = 1\,\mathrm{m^2/sec^2}$
		Old: rad	rad	$1\,\mathrm{rad} = 10^{-2}\,\mathrm{Gy}$
Dose equivalent	H	Sievert	Sv	$1\,\mathrm{J\,kg^{-1}} = 1\,\mathrm{m^2/sec^2}$
		Old: rem	rem	$1\,\mathrm{rem} = 10^{-2}\,\mathrm{Sv}$

anode, the latter being preferred for protein X-ray crystallography because of their higher intensity. The most commonly used radiation from a tube has a wavelength of 1.5418 Å. This is the characteristic K_α wavelength emitted by a copper anode and is selected from the spectral distribution by a filter or, preferably, with a graphite monochromator. The extremely high intensity X-ray radiation from a synchrotron is of great value for collecting data from weakly diffracting specimens. The beam is not only strong, it is also highly parallel, causing smaller but more brilliant spots on the detector. Therefore, with synchrotron radiation the resolution is somewhat better than in the home laboratory (more diffracted beams at high diffraction angle with greater details in the resulting protein structure).

Another advantage of synchrotron radiation is its tunability, which allows the selection of radiation with a wavelength at or below 1 Å. Although the specimen diffracts this radiation more weakly than the 1.5418 Å copper radiation, the fluorescent type detectors (image plate and Fast area detector) are more sensitive at this shorter wavelength. Another important advantage is the lower absorption of radiation with shorter wavelength and, consequently, less radiation damage to the crystals. For a protein structure determination the number of diffracted beams to be recorded is extremely high, of the order of $10^4–10^5$. To achieve this within a reasonable time requires highly efficient hardware, either an electronic area detector or an image plate. These instruments have completely changed protein X-ray crystallography by considerably reducing the time for exposure and data processing, and in this way have solved the previously time-consuming data collection problem. Of the more classical instruments the precession camera occupies a special place. Although the newer generation of protein crystallographers will not use it much if at all, it was the favorite instrument for the older generation. A new crystal form was always investigated first on the precession camera. From the undistorted image of the reciprocal lattice it presented, the cell dimensions could be easily measured, the symmetry in the crystal found, and the quality of the crystals determined. Moreover, crystals of heavy atom derivatives were qualitatively checked on the precession camera for their application in the structure determination by observing cell dimensions and possible intensity changes. A precession instrument with an image plate instead of a photographic film would be a very useful instrument in a protein X-ray crystallography laboratory.

Chapter 3
Crystals

3.1. Introduction

The beauty and regularity of crystals impressed people to such an extent that in the past crystals were regarded as products of nature with mysterious properties. Scientific investigation of crystals started in 1669, when Nicolaus Steno, a Dane working as a court physician in Tuscan, proposed, that during crystal growth *the angles between the faces remained constant.*

For a given crystal form individual crystals may differ in shape, that is, in the development of their faces, but they always have identical angles between the same faces (Figure 3.1). The specific morphology may depend on factors such as the supply of material during growth, on the presence of certain substances in the mother liquor, or on the mother liquor itself. For a single crystal form the angles between the faces are constant, but this is not true if the crystals belong to different crystal forms. Figure 3.2 shows four different crystal forms of deoxyhemoglobin from the sea lamprey *Petromyzon marinus.* Their appearance depends on the buffer and on the precipitating agent, although occasionally two different forms appear under the same conditions.

Before the famous first X-ray crystallographic diffraction experiment by von Laue, Friedrich, and Knipping in 1912, the internal regularity of a crystal was suggested but never proven. X-ray crystallography has dramatically changed this situation.

Determining the atomic structure of a molecule, particularly one as complex as a protein molecule, is greatly facilitated if a large number of identical molecules can be aggregated in a regular arrangement. The highest order is present in crystals of the material, although structural

54

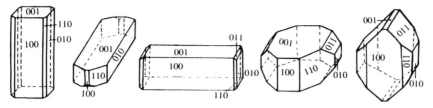

Figure 3.1. Crystals of trimethylammonium bromide belonging to the same crystal form but exhibiting a range of morphologies.

information can also be obtained from fibers. Not much can be done with an amorphous solid or a solution, which give weak and diffuse X-ray diffraction patterns and from which little structural information can be derived. In this book we restrict ourselves to crystals.

A crystal of organic material is a three-dimensional periodic arrangement of molecules. When the material precipitates from a solution, its molecules attempt to reach the lowest free energy state. This is often accomplished by packing them in a regular way; in other words, a crystal grows. It is surprising to observe that even large protein molecules follow this principle, although occasionally they unfortunately do not crystallize. Flat planes at the surface of a well-developed crystal reflect the regular packing of the molecules in the crystal. In this regular packing three repeating vectors $\mathbf{a}^1$, $\mathbf{b}$, and $\mathbf{c}$ can be recognized with the angles α, β, and γ between them. These three vectors define a unit cell in the crystal lattice (Figure 3.3).

If the content of the unit cells is neglected for the moment, the crystal can be regarded as a three-dimensional stack of unit cells with their edges forming a grid or lattice (Figure 3.4). The line in the $\mathbf{a}$ direction is called

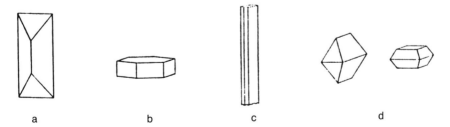

Figure 3.2. The crystals a–d belong to four different crystal forms of deoxyhemoglobin from the sea lamprey *Petromyzon marinus*. Reproduced with permission from Hendrickson et al. (1968).

---

[1] Vectors are in boldface type.

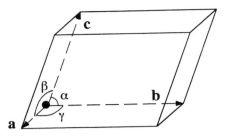

Figure 3.3. One unit cell in the crystal lattice.

● Origin

the x-axis of the lattice; the y-axis is in the **b** direction and the z-axis is in the **c** direction. The x-, y- and z-axes together form a coordinate system that by convention is right-handed. We shall later see (Section 4.7) that diffraction of X-rays by a crystal can be regarded as reflection against planes in the lattice. These planes are constructed through the lattice points and a great many sets of these planes can be drawn (Figure 3.5).

Within a set, the planes are parallel and equidistant with perpendicular distance d. As can be derived from Figure 3.5 the lattice planes cut an axis, for example the x-axis, into equal parts that have a length $a/1$, $a/2$, $a/3$, $a/4$ etc. The whole numbers 1, 2, 3, 4, ... are called indices. A set of lattice planes is determined by three indices h, k, and l, if the planes cut the x-axis in a/h, y in b/k, and z in c/l pieces. If a set of planes is parallel to an axis that particular index is 0 (the plane intercepts the axis at

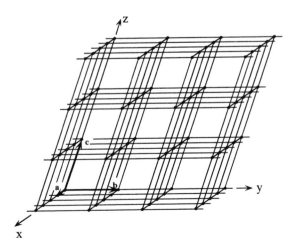

Figure 3.4. A crystal lattice is a three-dimensional stack of unit cells.

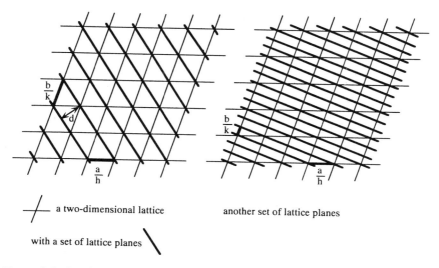

a two-dimensional lattice

with a set of lattice planes

another set of lattice planes

Figure 3.5. Lattice planes in a two-dimensional lattice.

infinity). Therefore, the unit cell is bounded by the planes (100), (010), and (001) (Figure 3.6). The flat faces of a crystal are always parallel to lattice planes (Figures 3.1 and 3.7). The parentheses in ($h\ k\ l$) are used to distinguish a lattice plane from a line segment in the unit cell, which is given in brackets: for example, [100] is the line segment from the origin of the unit cell to the end of the a-axis and [111] is the body diagonal from the origin to the opposite corner.

From Figure 3.5 it is clear that the projection of a/h, b/k, and c/l on the line perpendicular to the corresponding lattice plane ($h\ k\ l$) is equal to the lattice plane distance d. We have not yet discussed the choice of unit cell in the crystal. For example, in Figure 3.8 the choice could be either unit cell I, II, or III. Often the problem does not exist because of symmetry considerations in the crystal (Section 3.2). If the choice does

Figure 3.6. One unit cell bounded by the planes (100), (010), and (001). The directions along **a**, **b**, and **c** are indicated by [100], [010], and [001].

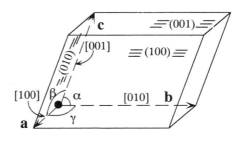

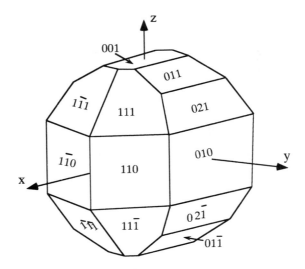

Figure 3.7. A crystal showing several faces.

exist then certain rules should be followed, given in *International Tables for Crystallography*, Vol. A, Chapter 9 (Hahn, 1983).

The main conditions are as follows:

1. The axis system should be right-handed.

2. The basis vectors should coincide as much as possible with directions of highest symmetry (Section 3.2).

3. The cell taken should be the smallest one that satisfies condition 2. This condition sometimes leads to the preference of a face-centered (A, B, C, or F) or a body-centered (I) cell over a primitive (P) smallest cell

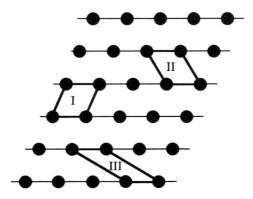

Figure 3.8. In this two-dimensional lattice the unit cell can be chosen in different ways: as I, as II, or as III.

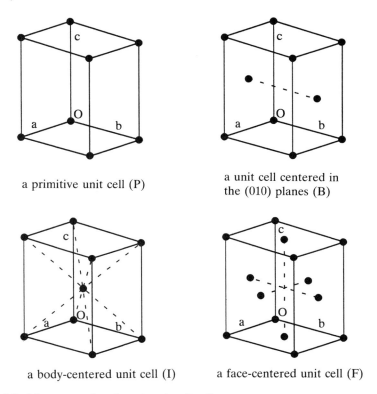

a primitive unit cell (P)

a unit cell centered in
the (010) planes (B)

a body-centered unit cell (I)   a face-centered unit cell (F)

Figure 3.9. Noncentered and centered unit cells.

(Figure 3.9). Primitive cells have only one lattice point per unit cell, whereas nonprimitive cells contain two or more lattice points per unit cell. They are designated A, B, or C if one of the faces of the cell is centered: it has extra lattice points on opposite faces of the unit cell, respectively, on the *bc* (A), *ac* (B), or *ab* (C) faces. If all faces are centered the designation is F (Figure 3.9).

4. Of all lattice vectors none is shorter than *a*.

5. Of those not directed along **a** none is shorter than *b*.

6. Of those not lying in the *a*, *b* plane none is shorter than *c*.

7. The three angles between the basis vectors **a**, **b** and **c** are either all acute ($<90°$) or all obtuse ($\geq90°$).

3.2. Symmetry

The search for a minimum free energy and as a consequence the regular packing of molecules in a crystal lattice often leads to a symmetric relationship between the molecules. As we have seen in the previous

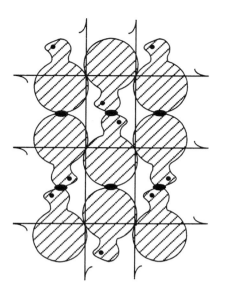

Figure 3.10. A two-dimensional lattice with 2-fold symmetry axes perpendicular to the plane of the figure and 2-fold screw axes in the plane.

section, a characteristic of a crystal is that it has unit translations in three dimensions, also called three-dimensional translational symmetry, corresponding to the repetition of the unit cells. Often, additional symmetry is encountered.

Examples are given in Figures 3.10 and 3.11, which show operations having 2- and 3-fold (screw) rotation axes as symmetry elements. Figures

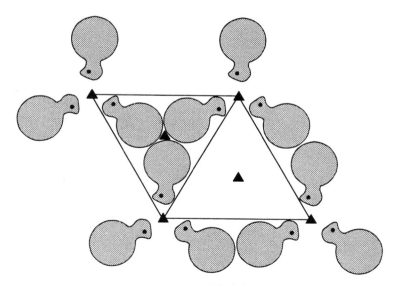

Figure 3.11. A two-dimensional lattice with 3-fold symmetry axes perpendicular to the plane of the figure.

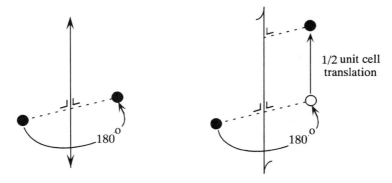

Figure 3.12. A 2-fold axis (left) and a 2-fold screw axis (right); the latter relates one molecule to another by a 180° rotation plus a translation over half of the unit cell.

3.12 and 3.13 give further examples of 2-fold and 3-fold symmetry operations. n-fold axes with $n = 5$ or $n > 6$ do not occur. The reason is that space cannot be filled completely with, for example, a 5-fold or a 7-fold axis. In addition to axes of symmetry, crystals can have mirror planes, inversion centers (centers of symmetry) (Figure 3.14), and rotation inversion axes, which combine an inversion and a rotation. Table 3.1 lists all possible symmetry operations together with their symbols and the translation operation.

Another way of looking at symmetry is the following. Application of the symmetry operators, such as rotations with or without translations,

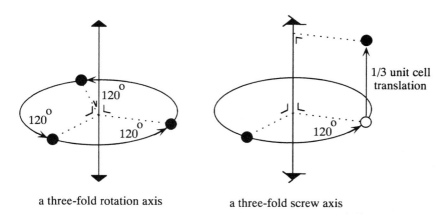

a three-fold rotation axis   a three-fold screw axis

Figure 3.13. A 3-fold axis (left) and a 3-fold screw axis (right); the latter relates one molecule to another by a 120° rotation and a translation over one-third of the unit cell.

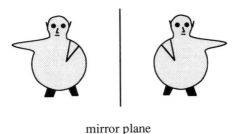

Figure 3.14. The effect of a mirror and of an inversion center.

mirror plane

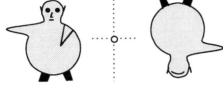

center of symmetry
or inversion center

leaves the entire crystal unchanged; it looks exactly as before. Therefore, the properties of the crystal, such as those of an electrical or mechanical nature, obey at least the same symmetry. There are 230 different ways to combine the allowed symmetry operations in a crystal, leading to 230 space groups. They can be found in the *International Tables of Crystallography*, Volume A (Hahn, 1983).

The restrictions given above for the symmetry axes of "ordinary" crystals are the consequence of their three-dimensional translation symmetry because this requires space to be filled entirely by identical repeating units. This symmetry is not present in so-called quasicrystals, which consist of two or more (but a finite number) different units. The position of the units in the lattice of a quasicrystal is determined according to a predictable sequence that never quite repeats: they are quasiperiodic and can have any rotational symmetry, including 5-fold and 7-fold, which are forbidden in "ordinary" crystals. So far no quasicrystalline protein crystals have been found and they will not be considered further.

3.3. Possible Symmetry for Protein Crystals

Not all 230 space groups are allowed for protein crystals. The reason is that in protein crystals the application of mirror planes and inversion centers (centers of symmetry) would change the asymmetry of the amino acids: an L-amino acid would become a D-amino acid, but these are never found in proteins. This limitation restricts the number of space groups for proteins appreciably: only those without any symmetry (triclinic) or with exclusively rotation or screw axes are allowed. However, mirror lines and inversion centers do occur in projections of protein structures along an axis. For example, a projection along a 2-fold axis has an inversion center and mirror lines do occur in a projection of the structure on a plane parallel to a 2-fold axis.

3.4. Coordinate Triplets: General and Special Positions

The position of a point P in the unit cell is given by its position vector $\mathbf{r}$ (Figure 3.15). In terms of its relative coordinates x, y, and z with respect to the crystal axes $\mathbf{a}$, $\mathbf{b}$, and $\mathbf{c}$, $\mathbf{r}$ is given by

$$\mathbf{r} = \mathbf{a}x + \mathbf{b}y + \mathbf{c}z \qquad (3.1)$$

The position of P can thus be described by its relative coordinates, that is, by its coordinate triplet x, y, z. The coordinate triplets of the points P and P', related by the 2-fold axis along $\mathbf{c}$ in Figure 3.15, are x, y, z and $-x$, $-y$, z. If a molecule occupies position x, y, z then an identical molecule occupies position $-x$, $-y$, z. These molecules are said to occupy general positions.

If, however, the molecule itself has a 2-fold axis that coincides with the crystallographic 2-fold axis, half of the molecule is mapped onto the other half by the symmetry operation. This molecule occupies a "special" position. Each special position has a certain point symmetry. In the present example the point symmetry would be given by the symbol 2.

3.5. Asymmetric Unit

If the lattice has a level of symmetry higher than triclinic, then each particle in the cell will be repeated a number of times as a consequence of the symmetry operations. For example in space group $P2_12_12_1$ (space group number 19 in the *International Tables*) one can always expect at least four equal particles in the unit cell related by the symmetry operations (Figure 3.16). This unit cell has four asymmetric units. The number of molecules in a unit cell is not necessarily equal to the number of asymmetric units. There may be two or more independent molecules in each asymmetric unit. On the other hand if a molecule occupies a special position, e.g., if a symmetry axis passes through a molecule, relating one

Table 3.1. Graphic Symbols for Symmetry Elements[a]

Symmetry axis or symmetry point	Graphic symbol	Screw vector of a right-handed screw rotation in units of the shortest lattice translation vector parallel to the axis	Printed symbol
Symmetry axes normal to the plane of projection (three dimensions) and symmetry points in the plane of the figure (two dimensions)			
Identity	None	None	1
Twofold rotation axis Twofold rotation point (two dimentions)	●	None	2
Twofold screw axis: "2 sub 1"	◗	$\frac{1}{2}$	2_1
Threefold rotation axis Threefold rotation point (two dimentions)	▲	None	3
Threefold screw axis: "3 sub 1"	◣	$\frac{1}{3}$	3_1
Three fold screw axis: "3 sub 2"	◢	$\frac{2}{3}$	3_2
Fourfold rotation axis Fourfold rotation point (two dimensions)	◆	None	4
Fourfold screw axis: "4 sub 1"	✦	$\frac{1}{4}$	4_1
Fourfold screw axis: "4 sub 2"	✦	$\frac{1}{2}$	4_2
Fourfold screw axis: "4 sub 3"	✦	$\frac{3}{4}$	4_3
Sixfold rotation axis Sixfold rotation point (two dimensions)	⬡	None	6

Description		Fraction	Notation
Sixfold screw axis: "6 sub 1"		$\frac{1}{6}$	6_1
Sixfold screw axis: "6 sub 2"		$\frac{1}{3}$	6_2
Sixfold screw axis: "6 sub 3"		$\frac{1}{2}$	6_3
Sixfold screw axis: "6 sub 4"		$\frac{2}{3}$	6_4
Sixfold screw axis: "6 sub 5"		$\frac{5}{6}$	6_5
Center of symmetry, inversion center: "1 bar" / Reflection point, mirror point (one dimension)	∘	None	$\bar{1}$
Twofold rotation axis with center of symmetry		None	$2/m$
Twofold screw axis with centre of symmetry		$\frac{1}{2}$	$2_1/m$
Inversion axis: "3 bar"		None	$\bar{3}$
Inversion axis: "4 bar"		None	$\bar{4}$
Fourfold rotation axis with center of symmetry		None	$4/m$
"4 sub 2" screw axis with center of symmetry		$\frac{1}{2}$	$4_2/m$
Inversion axis: "6 bar"		None	$\bar{6}$
Sixfold rotation axis with center of symmetry		None	$6/m$
"6 sub 3" screw axis with center of symmetry		$\frac{1}{2}$	$6_3/m$

(continued)

Table 3.1. *Continued*

Symmetry axis or symmetry point	Graphic symbol	Screw vector of a right-handed screw rotation in units of the shortest lattice translation vector parallel to the axis	Printed symbol
Symmetry axes parallel to the plane of projection			
Twofold rotation axis		None	2
Twofold screw axis: "2 sub 1"		$\frac{1}{2}$	2_1
Fourfold rotation axis		None	4
Fourfold screw axis: "4 sub 1"		$\frac{1}{4}$	4_1
Fourfold screw axis: "4 sub 2"		$\frac{1}{2}$	4_2
Fourfold screw axis: "4 sub 3"		$\frac{3}{4}$	4_3
Inversion axis: "4 bar"		None	$\bar{4}$
Symmetry axes inclined to the plane of projection (in cubic space groups only)			
Twofold rotation axis		None	2
Twofold screw axis: "2 sub 1"		$\frac{1}{2}$	2_1
Threefold rotation axis		None	3
Threefold screw axis: "3 sub 1"		$\frac{1}{3}$	3_1
Threefold screw axis: "3 sub 2"		$\frac{2}{3}$	3_2
Inversion axis: "3 bar"		None	$\bar{3}$

[a] Reprinted from the *International Tables of Crystallography*, Volume A (Hahn, 1983), with permission of The International Union of Crystallography.

Figure 3.15. This crystal has a 2-fold axis along c. The point P with coordinate triplet x, y, z, is related by the symmetry operation to point P' with coordinate triplet $-x, -y, z$.

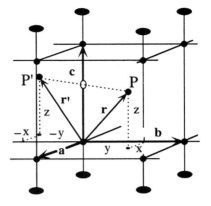

part of the molecule to one or more other parts in that molecule, the unit cell contains fewer molecules than anticipated from the number of asymmetric units.

It is important to note that molecules related by crystallographic symmetry are identical and have identical crystallographic environments. However, if two or more molecules occur in the asymmetric unit they do not have an identical environment and, moreover, they may differ in conformation.

3.6. Point Groups

A characteristic feature of crystals is the presence of flat boundary faces with sharp edges between them. Internal symmetry in the crystal is reflected in the arrangement of these boundary faces. However, the

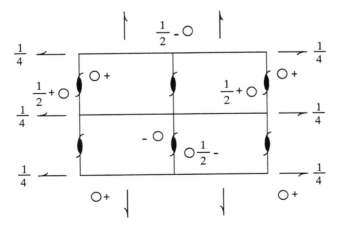

Figure 3.16. The projection of a $P2_12_12_1$ unit cell; it contains four asymmetric units.

translation components of the symmetry operations cannot be observed
on the macroscopic level and for the outer shape of the crystals we are
left with symmetry without translation, e.g., a 2-fold screw axis becomes
a 2-fold axis. This limited group of symmetry elements forms a collection
of point groups, because the symmetry elements always pass through one
point. The absence of 5-fold, 7-fold etc. axes in crystals (because of the
requirement of unit translation symmetry) limits the number of point
groups to 32. The absence of a 5-fold crystallographic axis does not mean
that a 5-fold axis never occurs. For instance, in virus particles no unit
translation symmetry exists and 5-fold axes do occur.

3.7. Crystal Systems

With the choice of the unit cell according to the rules in Section 3.1, the
32 point groups can be assigned to seven and not more than seven
crystallographic systems, shown in Table 3.2. This limitation to seven
systems is due to the combination of symmetry elements. We shall show
this with an example.

If the unit cell has one 2-fold axis, the system is clearly monoclinic. If a
second 2-fold axis is added, the two axes must be perpendicular or make
an angle of 30° or 60° with each other, because otherwise an unlimited
number of 2-fold axes would be generated in the same plane (Figure
3.17). From this figure it is evident that two perpendicular 2-fold axes
generate a third one perpendicular to the plane of the first two axes.
Therefore, crystal symmetry excludes the existence of a crystal system
with one angle equal to 90° and two angles different from 90°.

Table 3.2. The Seven Crystal Systems

Crystal system	Conditions imposed on cell geometry	Minimum point group symmetry
Triclinic	None	1
Monoclinic	$\alpha = \gamma = 90°$ (b is the unique axis; for proteins this is a 2-fold axis or screw axis)	2
	or: $\alpha = \beta = 90°$ (c is unique axis; for proteins this is a 2-fold axis or screw axis)	
Orthorhombic	$\alpha = \beta = \gamma = 90°$	222
Tetragonal	$a = b; \alpha = \beta = \gamma = 90°$	4
Trigonal	$a = b; \alpha = \beta = 90°; \gamma = 120°$ (hexagonal axes)	3
	or: $a = b = c; \alpha = \beta = \gamma$ (rhombohedral axes)	
Hexagonal	$a = b; \alpha = \beta = 90°; \gamma = 120°$	6
Cubic	$a = b = c; \alpha = \beta = \gamma = 90°$	23

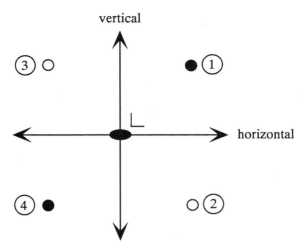

Figure 3.17. The two 2-fold axes in the plane of the page relate the four circles to each other. The open circles are above the plane, the filled circles below it. If a molecule is placed at position 1, the horizontal 2-fold axis generates a molecule in position 2 and the vertical axis in position 3. But then the horizontal axis also generates a molecule in position 4. It is evident that a third 2-fold axis is generated perpendicular to the first two.

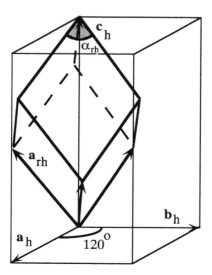

Figure 3.18. A rhombohedral unit cell and its corresponding hexagonal cell.

The trigonal system can be treated either with hexagonal axes or with rhombohedral axes (Figure 3.18). In the hexagonal unit cell a and b are equal in length and have an angle of 120° to each other. c is perpendicular to the ab-plane and differs in length from a and b. The rhombohedral cell has three equal axes at angles not necessarily 90° with each other. It can be regarded as a cube either compressed or elongated along a body diagonal. The rhombohedral cell corresponds to a hexagonal cell centered at 2/3, 1/3, 1/3 and 1/3, 2/3, 2/3. The relation between the cell parameters a and c of the hexagonal cell and the parameters a' and α' of the rhombohedral cell are as follows:

$$a = a'\sqrt{2}\sqrt{1 - \cos \alpha'} = 2a' \sin \frac{\alpha'}{2}$$

$$c = a'\sqrt{3}\sqrt{1 + 2\cos \alpha'}$$

$$\frac{c}{a} = \sqrt{\frac{3}{2}}\sqrt{\frac{1 + 2\cos\alpha'}{1 - \cos\alpha'}} = \sqrt{\frac{9}{4\sin^2(\alpha'/2)} - 3}$$

$$a' = \frac{1}{3}\sqrt{3a^2 + c^2}$$

$$\sin\frac{\alpha'}{2} = \frac{3}{2\sqrt{3 + (c^2/a^2)}} \quad \text{or} \quad \cos\alpha' = \frac{(c^2/a^2) - 3/2}{(c^2/a^2) + 3}$$

It is important to realize that the conditions imposed on cell geometry are not sufficient to distinguish between the crystal systems. For instance, if a unit cell is found to have three angles of 90° it does not necessarily mean that the crystal belongs to the orthorhombic system. It could also be triclinic with three angles of 90° in the unit cell by coincidence. By being orthorhombic it is in addition required to have a minimum point group symmetry of 222 in the crystal, which expresses itself as mmm symmetry in the diffraction pattern ("mmm" means three perpendicular mirror planes).

3.8. Radiation Damage

In Chapter 2 it was mentioned that the high energy photons of X-rays have a harmful effect on living tissue. This also applies to protein crystals, which undergo radiation damage if exposed to X-rays. The high energy X-ray photons cause the formation of radicals in the crystal, which leads to subsequent chemical reactions, that gradually destroy the crystalline order. Radiation damage is sometimes so serious that after a few hours of exposure (on a rotating anode tube) the X-ray pattern dies away. However, normally a crystal can be exposed for 100 hr or more before it must be replaced. With modern, sensitive, X-ray detectors exposure times can be much shorter than with the traditional detectors like film or scintillation counters, and a complete data set can often be obtained from one crystal. Cooling the crystals slows down the destructive process appreciably. Some protein crystals can—with some precautions—even be cooled to liquid nitrogen temperature. At that temperature radiation damage virtually disappears.

3.9. Characterization of the Crystals

If crystals of sufficient size (≥ 0.2 mm) have been grown, they must first be characterized:

- What is their quality?
- What are the unit cell dimensions?
- To what space group do they belong?
- How many protein molecules are in the unit cell and in one asymmetric unit?

The quality of the crystals depends on the ordering of the molecules in the unit cell. Because of thermal vibrations and static disorder, the positions of the atoms are not strictly fixed. As a consequence the intensities of the X-ray reflections drop at higher diffraction angles. In Section 4.7 it will be shown that the diffracted X-ray beams can be considered as being reflected against lattice planes, and that the relation between the lattice plane distance d and the diffraction angle θ is given by $2d \sin \theta = \lambda$ (Bragg's law). Diffraction patterns with maximal observed resolution corresponding to a lattice spacing of 5 Å can be regarded as poor, of 2.0–2.5 Å as normal, and of 1.0–1.5 Å as very high.

Most crystals cannot be considered as ideal single crystals, because the regular repetition of the unit cells is interrupted by lattice defects. The diffraction pattern of such crystals can be regarded as the sum of the diffraction patterns originating from mosaic blocks with slightly different orientations. The mosaicity in good quality protein crystals is moderate, between 0.2 and 0.5 degrees. The cell dimensions can easily be derived from the diffraction pattern collected on X-ray film, image plate, or measured with a diffractometer or area detector.

From the symmetry in the diffraction pattern and the systematic absence of specific reflections in that pattern, one can deduce the space group to which the crystal belongs. As discussed in Section 3.3, space groups with mirror planes or inversion centers do not apply to protein crystals. An inversion center does occur, however, *in the X-ray diffraction pattern* if anomalous scattering (Section 7.8) is absent.

An estimation of the number of molecules per unit cell (Z) can be made by a method proposed by Matthews. He found that for many protein crystals the ratio of the unit cell volume and the molecular weight is between 1.7 and 3.5 Å^3/Da with most values around 2.15 Å^3/Da. This number is called V_M (Matthews, 1968).

Example: Suppose a crystal belongs to space group $C2$ and has a unit cell volume of $319,000\,\text{Å}^3$. The molecular weight M_r of the protein is known to be $32,100$. Then, for Z of 2, 4, or 8, V_M is, respectively, 5, 2.5, or $1.25\,\text{Å}^3/\text{Da}$. This crystal most likely has four molecules in the unit cell. Space group $C2$ (no. 5 in the *International Tables*) has four asymmetric units and, therefore, there is one protein molecule per asymmetric unit.

For a more accurate determination of the number of protein molecules per unit cell, the density of the crystal and its solvent content should be determined. The density can best be measured in a density gradient column, made up from organic solvents, or from a water solution of the polymer compound Ficoll in a concentration of 30–60%. Because of the high viscosity of the Ficoll solutions, these measurements must be done in a centrifuge.

The solvent content of the crystals could—in principle—be determined by weighing a number of crystals before and after drying. However, a problem arises because in the wet state, solvent around the crystals must be removed, but not internal solvent that fills the pores inside the crystals and that is part of the crystal structure. This separation is difficult to achieve. With the unit cell volume (V_{cell}), the density ρ, and the solvent content (weight fraction x) known, Z can be calculated:

$$\frac{Z \times M_r}{N} = (1 - x) \times \rho \times V_{cell} \tag{3.2}$$

where N is Avogadro's number. From V_M, the volume fraction of the solvent in the crystal can be calculated in the following way:

$$
\begin{aligned}
V_{protein} &= \frac{\text{Volume of protein in the unit cell}}{V_{cell}} \\
&= \frac{(Z \times M_r \times \text{specific volume of the protein})/N}{V_M \times Z \times M_r} \\
&= \frac{\text{specific volume in cm}^3/\text{g}}{V_M \text{ in Å}^3/\text{Da} \times N \text{ mol}^{-1}}
\end{aligned}
$$

The specific volume of a protein molecule is always approximately $0.74\,\text{cm}^3/\text{g}$, and this gives

$$V_{protein} = 1.23/V_M \quad \text{and} \quad V_{solvent} = 1 - 1.23/V_M.$$

For the example given above, where $V_M = 2.5$, $V_{solvent} = 0.49$.

Summary

People have been fascinated by crystals since prehistoric times. However, before the introduction of X-ray diffraction in 1912 only their external properties could be studied, such as the angles between their faces or